# IMPROVING
# DISASTER MANAGEMENT

THE ROLE OF IT IN MITIGATION, PREPAREDNESS,
RESPONSE, AND RECOVERY

Ramesh R. Rao, Jon Eisenberg, and Ted Schmitt, *Editors*

Committee on Using Information Technology to
Enhance Disaster Management

Computer Science and Telecommunications Board

Division on Engineering and Physical Sciences

**NATIONAL RESEARCH COUNCIL**
*OF THE NATIONAL ACADEMIES*

THE NATIONAL ACADEMIES PRESS
Washington, D.C.
**www.nap.edu**

THE NATIONAL ACADEMIES PRESS  500 Fifth Street, N.W.  Washington, DC 20001

NOTICE: The project that is the subject of this report was approved by the Governing Board of the National Research Council, whose members are drawn from the councils of the National Academy of Sciences, the National Academy of Engineering, and the Institute of Medicine. The members of the committee responsible for the report were chosen for their special competences and with regard for appropriate balance.

This study was supported by the Battelle Memorial Institute under subcontract number 189936 to a contract between the Federal Emergency Management Agency and the Battelle Memorial Institute. Any opinions, findings, conclusions, or recommendations expressed in this publication are those of the authors and do not necessarily reflect the views of the organizations and agencies that provided support for the project.

International Standard Book Number-13: 978-0-309-10396-1
International Standard Book Number-10: 0-309-10396-7

Additional copies of this report are available from the National Academies Press, 500 Fifth Street, N.W., Lockbox 285, Washington, DC 20055; (800) 624-6242 or (202) 334-3313 (in the Washington metropolitan area); Internet, http://www.nap.edu.

Copyright 2007 by the National Academy of Sciences. All rights reserved.

Printed in the United States of America

# THE NATIONAL ACADEMIES
*Advisers to the Nation on Science, Engineering, and Medicine*

The **National Academy of Sciences** is a private, nonprofit, self-perpetuating society of distinguished scholars engaged in scientific and engineering research, dedicated to the furtherance of science and technology and to their use for the general welfare. Upon the authority of the charter granted to it by the Congress in 1863, the Academy has a mandate that requires it to advise the federal government on scientific and technical matters. Dr. Ralph J. Cicerone is president of the National Academy of Sciences.

The **National Academy of Engineering** was established in 1964, under the charter of the National Academy of Sciences, as a parallel organization of outstanding engineers. It is autonomous in its administration and in the selection of its members, sharing with the National Academy of Sciences the responsibility for advising the federal government. The National Academy of Engineering also sponsors engineering programs aimed at meeting national needs, encourages education and research, and recognizes the superior achievements of engineers. Dr. Wm. A. Wulf is president of the National Academy of Engineering.

The **Institute of Medicine** was established in 1970 by the National Academy of Sciences to secure the services of eminent members of appropriate professions in the examination of policy matters pertaining to the health of the public. The Institute acts under the responsibility given to the National Academy of Sciences by its congressional charter to be an adviser to the federal government and, upon its own initiative, to identify issues of medical care, research, and education. Dr. Harvey V. Fineberg is president of the Institute of Medicine.

The **National Research Council** was organized by the National Academy of Sciences in 1916 to associate the broad community of science and technology with the Academy's purposes of furthering knowledge and advising the federal government. Functioning in accordance with general policies determined by the Academy, the Council has become the principal operating agency of both the National Academy of Sciences and the National Academy of Engineering in providing services to the government, the public, and the scientific and engineering communities. The Council is administered jointly by both Academies and the Institute of Medicine. Dr. Ralph J. Cicerone and Dr. Wm. A. Wulf are chair and vice chair, respectively, of the National Research Council.

**www.national-academies.org**

## COMMITTEE ON USING INFORMATION TECHNOLOGY TO ENHANCE DISASTER MANAGEMENT

RAMESH R. RAO, University of California, San Diego, *Chair*
YIGAL ARENS, University of Southern California
ART BOTTERELL, Contra Costa County, California, Office of the Sheriff
TIMOTHY X. BROWN, University of Colorado, Boulder
JOHN R. HARRALD, George Washington University
RICHARD HOWARD, Rutgers University
NANCY JESUALE, NetCity Engineering, Inc.
DAVID KEHRLEIN, Environmental Science Research Institute
WILLIAM MAHEU, San Diego, California, Police Department
ROBIN R. MURPHY, University of South Florida
ROBERT NECHES, University of Southern California
MASANOBU SHINOZUKA, University of California, Irvine
ELLIS STANLEY, City of Los Angeles
PETER STEENKISTE, Carnegie Mellon University
GIO WIEDERHOLD, Stanford University

**Staff**

JON EISENBERG, Study Director
TED SCHMITT, Program Officer
DAVID PADGHAM, Associate Program Officer
GLORIA WESTBROOK, Senior Program Assistant (through December 2006)
JENNIFER M. BISHOP, Program Associate

# COMPUTER SCIENCE AND TELECOMMUNICATIONS BOARD

JOSEPH F. TRAUB, Columbia University, *Chair*
ERIC BENHAMOU, Benhamou Global Ventures, LLC
FREDERICK R. CHANG, University of Texas, Austin
WILLIAM DALLY, Stanford University
MARK E. DEAN, IBM Almaden Research Center
DAVID J. DEWITT, University of Wisconsin, Madison
DEBORAH ESTRIN, University of California, Los Angeles
JOAN FEIGENBAUM, Yale University
KEVIN KAHN, Intel Corporation
JAMES KAJIYA, Microsoft Corporation
MICHAEL KATZ, University of California, Berkeley
RANDY H. KATZ, University of California, Berkeley
SARA KIESLER, Carnegie Mellon University
TERESA H. MENG, Stanford University
PRABHAKAR RAGHAVAN, Yahoo! Research
FRED B. SCHNEIDER, Cornell University
ALFRED Z. SPECTOR, Independent Consultant, Pelham, New York
WILLIAM STEAD, Vanderbilt University
ANDREW J. VITERBI, Viterbi Group, LLC
PETER WEINBERGER, Google, Inc.
JEANNETTE M. WING, Carnegie Mellon University

JON EISENBERG, Director
KRISTEN BATCH, Associate Program Officer
RADHIKA CHARI, Administrative Coordinator
RENEE HAWKINS, Financial Associate
MARGARET MARSH HUYNH, Senior Program Assistant
HERBERT S. LIN, Senior Scientist
LYNETTE I. MILLETT, Senior Program Officer
DAVID PADGHAM, Associate Program Officer
JANICE SABUDA, Senior Program Assistant
TED SCHMITT, Program Officer
BRANDYE WILLIAMS, Program Assistant
JOAN WINSTON, Program Officer


For more information on CSTB, see its Web site at http://www.cstb.org, write to CSTB, National Research Council, 500 Fifth Street, N.W., Washington, DC 20001, call (202) 334-2605, or e-mail at cstb@nas.edu.

# Preface

To improve how information technology is used in disaster management, Section 214 of the E-Government Act of 2002 called on the administrator of the Office of Electronic Government in the Office of Management and Budget, in consultation with the Federal Emergency Management Agency (FEMA), to "ensure that a study is conducted on using information technology to enhance crisis preparedness, response, and consequence management of natural and manmade disasters" (see Box P.1). In early 2005, in response to a request from FEMA to the National Research Council (NRC), via a contract with Battelle Memorial Institute, the Committee on Using Information Technology to Enhance Disaster Management was established under the auspices of the Computer Science and Telecommunications Board to study these issues. The committee's first report, *Summary of a Workshop on Using Information Technology to Enhance Disaster Management*,[1] summarized the discussions at a public workshop held on June 22-23, 2005. Representatives of federal, state, and local government agencies; private industry; and the research community participated in the workshop.

Over the next year the committee met four times and made several site visits to gather input from federal agencies; state and local public

---

[1] National Research Council, *Summary of a Workshop on Using Information Technology to Enhance Disaster Management*, The National Academies Press, Washington, D.C., September 2005.

### BOX P.1
### Section 214 of the E-Government Act of 2002, Public Law 107-347

SEC. 214. ENHANCING CRISIS MANAGEMENT THROUGH ADVANCED INFORMATION TECHNOLOGY

(a) PURPOSE.—The purpose of this section is to improve how information technology is used in coordinating and facilitating information on disaster preparedness, response, and recovery, while ensuring the availability of such information across multiple access channels.

(b) IN GENERAL.—

(1) STUDY ON ENHANCEMENT OF CRISIS RESPONSE.—Not later than 90 days after the date of enactment of this Act, the Administrator, in consultation with the Federal Emergency Management Agency, shall ensure that a study is conducted on using information technology to enhance crisis preparedness, response, and consequence management of natural and manmade disasters.

(2) CONTENTS.—The study under this subsection shall address—

(A) a research and implementation strategy for effective use of information technology in crisis response and consequence management, including the more effective use of technologies, management of information technology research initiatives, and incorporation of research advances into the information and communications systems of—

(i) the Federal Emergency Management Agency; and

(ii) other Federal, State, and local agencies responsible for crisis preparedness, response, and consequence management; and

(B) opportunities for research and development on enhanced technologies into areas of potential improvement as determined during the course of the study.

(3) REPORT.—Not later than 2 years after the date on which a contract is entered into under paragraph (1), the Administrator shall submit a report on the study, including findings and recommendations to—

(A) the Committee on Governmental Affairs of the Senate; and

(B) the Committee on Government Reform of the House of Representatives.

(4) INTERAGENCY COOPERATION.—Other Federal departments and agencies with responsibility for disaster relief and emergency assistance shall fully cooperate with the Administrator in carrying out this section.

(5) AUTHORIZATION OF APPROPRIATIONS.—There are authorized to be appropriated for research under this subsection, such sums as are necessary for fiscal year 2003.

(c) PILOT PROJECTS.—Based on the results of the research conducted under subsection (b), the Administrator, in consultation with the Federal Emergency Management Agency, shall initiate pilot projects or report to Congress on other activities that further the goal of maximizing the utility of information technology in disaster management. The Administrator shall cooperate with other relevant agencies, and, if appropriate, State, local, and tribal governments, in initiating such pilot projects.

safety officials and emergency management practitioners; experts in disaster management; information technology researchers; and hardware and software vendors. In preparing this, its final report, the committee also drew on perspectives and information gleaned from professional conferences, the technical literature, and government reports.

Chapter 1 briefly characterizes disaster management, placing the use of information and communication technology in the broader human and organizational context and providing a framework for considering the range and nature of information and communication needs. Chapter 2 presents the committee's vision of the potential for information and communication technology to improve disaster management. Chapter 3 focuses on structural, organizational, and other non-technical barriers to the acquisition, adoption, and effective use of IT in disaster management. Chapter 4 provides an initial outline of the elements of a research program aimed at strengthening IT-enabled capabilities for disaster management.

During the development of this report, Hurricane Katrina struck the U.S. Gulf Coast. In the days following the hurricane's landfall, damage to the communications infrastructure, together with a host of other communications and information concerns, was cited by decision makers and reported on in the press as among the major challenges facing those involved in response and recovery efforts. The tragic events that occurred in Katrina's wake have, of course, served to underscore the importance of disaster management; they have also highlighted the role of information technology in disaster management, the interplay between technical and organizational considerations, and the contributions that research and development in these areas could make to future disaster management activities. However, although a number of the inputs focused on Katrina, the committee's charge, its deliberations, and this report encompass disasters in all (natural and human-made) forms and in all phases, from mitigation and preparedness to response and recovery.

> Ramesh R. Rao, *Chair*
> Committee on Using Information Technology
> to Enhance Disaster Management

# Acknowledgment of Reviewers

This report has been reviewed in draft form by individuals chosen for their diverse perspectives and technical expertise, in accordance with procedures approved by the National Research Council's Report Review Committee. The purpose of this independent review is to provide candid and critical comments that will assist the institution in making its published report as sound as possible and to ensure that the report meets institutional standards for objectivity, evidence, and responsiveness to the study charge. The review comments and draft manuscript remain confidential to protect the integrity of the deliberative process. We wish to thank the following individuals for their review of this report:

David Borth, Motorola,
Thomas J. Cowper, New York State Police,
Sharon Dawes, University at Albany, State University of New York,
Otto Doll, State of South Dakota Bureau of Information and Telecommunications,
Mica Endsley, SA Technologies,
Al Flax, Consultant, Potomac, Maryland,
W. Craig Fugate, State of Florida Office of Emergency Management,
Sara Kiesler, Carnegie Mellon University,
Prabhakar Ragahvan, Yahoo!,
Eric Rasmussen, U.S. Navy Medical Corps,
Myra Socher, TriMed, Inc.,
Kathleen Tierney, University of Colorado, Boulder, and
Charles Werner, Charlottesville, Virginia, Fire Department.

Although the reviewers listed above provided many constructive comments and suggestions, they were not asked to endorse the conclusions or recommendations, nor did they see the final draft of the report before its release. The review of this report was overseen by Samuel H. Fuller, Analog Devices, Inc., and Richard N. Wright, National Institute of Standards and Technology (retired). Appointed by the National Research Council, they were responsible for making certain that an independent examination of this report was carried out in accordance with institutional procedures and that all review comments were carefully considered. Responsibility for the final content of this report rests entirely with the authoring committee and the institution.

# Contents

SUMMARY AND RECOMMENDATIONS 1

1 INTRODUCTION AND CONTEXT 15

2 THE POTENTIAL TO ENHANCE DISASTER MANAGEMENT: KEY IT-BASED CAPABILITIES 34

3 IMPROVING ACQUISITION AND ADOPTION OF IT FOR DISASTER MANAGEMENT 68

4 ELEMENTS OF A RESEARCH AGENDA 108

APPENDIXES

A Illustrative Fictional Narratives of IT Use in Disaster Management 135
B Review of Interoperability Initiatives 151
C Workshop Agenda 157
D Speakers and Participants at Meetings and Site Visits 165
E Biographies of Committee Members and Staff 167

# Summary and Recommendations

Section 214 of the E-Government Act of 2002 called on the Office of Management and Budget, in consultation with the Federal Emergency Management Agency (FEMA), to "ensure that a study is conducted on using information technology to enhance crisis preparedness, response, and consequence management of natural and manmade disasters." This, the final report from the National Research Council's Committee on Using Information Technology to Enhance Disaster Management, addresses the issues listed in Section 214 and provides recommendations for enhancing disaster management through the use of IT.[1] In this study, *disasters* are defined as natural, technological, and human-initiated events that disrupt the normal functioning of the economy and society on a large scale; *information technology* (IT) is broadly defined as including computing and communications technology; and *disaster management* is defined as encompassing mitigation, preparedness, response, and recovery efforts undertaken to reduce the impact of disasters.

The purpose of this report is to inform federal, state, and local policy makers and public safety and emergency management professionals about future opportunities for the application of IT to disaster manage-

---

[1]The committee's first report—National Research Council, *Summary of a Workshop on Using Information Technology to Enhance Disaster Management*, The National Academies Press, Washington, D.C.—was published in September 2005.

ment. Many of the report's recommendations are aimed at the diverse set of federal, state, and local agencies and other organizations (referred to here as disaster management organizations) with responsibility for disaster management activities. Several recommendations indicate what might be done at the federal level to foster IT innovation that would enhance disaster management efforts, but it was beyond the scope of this study to recommend exactly where responsibility for implementing these recommendations should be placed within the federal government. This report is not intended as a comprehensive look at the complex, highly multidisciplinary topic of disaster management, nor does it explicitly address tradeoffs between investments in IT and other capabilities for disaster management or make recommendations about levels of funding for IT (or indeed other) disaster management activities.

## USING INFORMATION TECHNOLOGY AS A POINT OF LEVERAGE TO ENHANCE DISASTER MANAGEMENT

The challenge of disaster management is reducing the harm disasters cause to society, the economy, and the lives of individuals and communities. That task requires disaster managers to reduce uncertainty, to calculate and compare costs and benefits, and to manage resources, often on a much larger scale and at a much faster pace than are supported by methods and means for solving ordinary problems. IT provides capabilities that can help people grasp the dynamic realities of a disaster more clearly and help them formulate better decisions more quickly. And IT can help keep better track of the myriad details involved in all phases of disaster management.

The committee concluded that IT has as-yet-unrealized potential to improve how communities, the nation, and the global community handle disasters. Briefings to the committee suggested that some progress is being made in using IT to enhance disaster management. Presentations made at its June 2005 workshop, additional briefings to the committee, and reports on responses to recent disasters indicated, however, that disaster management organizations have not fully exploited many of today's technology opportunities. This situation stands in contrast to the considerable success enjoyed by some sectors such as financial services and transportation in adopting new IT technologies routinely and aggressively.

This report describes both short- and long-term opportunities to enhance responsiveness and increase resilience by applying IT to disaster management. As in other sectors, successful use of IT involves multiple factors—making smarter use of existing technologies, creating opportunities to develop and adopt new technologies, and evolving organizational practices to best employ those technologies. Accordingly, this report also examines mechanisms to facilitate the development and effective use of IT.

## Short-Term Opportunities to Use IT

Although the committee believes that investment in IT research and development (R&D) for disaster management should be guided in the long run by a comprehensive, stakeholder-driven roadmap (see below), it also sees opportunities for short-term investment in a number of specific areas that would yield significant benefits. The committee heard of many instances in which responders were able to make use of readily available technology—either provided by their organizations or acquired personally—that proved valuable during a disaster. The network effects associated with many of these technologies can create a critical mass of users that provides a potential point of interoperability and cooperation across agencies. For example, ad hoc use of 802.11x wireless capabilities in laptops carried by some first responders, peer-to-peer use of Land Mobile Radio System (LMRS) radios, and use of Family Radio Service/General Mobile Radio Service "walkie-talkies" all can help to provide communications even when the communications infrastructure is damaged. Such technology options may already be in the hands of users but may not be deployed in disasters because policies and procedures for their use are not in place.

Other examples of "low-hanging fruit" include the following:

- Use of sensors, wikis (editable Web sites), blogs, and data-mining tools to capture, analyze, and share lessons learned from operational experiences;
- Use of database, Web, and call center technologies to establish a service to provide information about available equipment, materiel, volunteers, and volunteer organizations;
- Use of planning, scheduling, task allocation, and resource management tools to help in formulating disaster management plans and tracking execution of the plans and to ensure timely recognition of problems and associated follow-up decision making; and
- Use of deployable cell phone technology to rapidly establish stand-alone communications capabilities for use in disasters where local infrastructure is damaged.

To exploit such short-term opportunities involves identifying them, establishing policies and procedures for their use, and providing training to users.

Recommendation 1: Disaster management organizations should take advantage of opportunities for adoption of existing technology or adjustment of policies and procedures that would allow significant short-term enhancement of disaster management.

### Key IT-Enabled Capabilities and R&D to Achieve Them

Making good decisions and taking appropriate action during extreme events require having access to communications, data, and computational resources that can be used to effectively coordinate a large number of geographically dispersed participants and assets, to exchange a wide variety of types of information, and to evaluate many scenarios and responses—all of which are changing dynamically.

The committee identified six key areas of IT-enabled capability (described in Box S.1 and discussed in more detail in Chapter 2) in which shorter-term development and longer-term research offer the potential for significant benefits:

- More robust, interoperable, and priority-sensitive communications;
- Better situational awareness and a common operating picture;
- Improved decision support and resource tracking and allocation;
- Greater organizational agility for disaster management;
- Better engagement of the public; and
- Enhanced infrastructure survivability and continuity of societal functions.

Some of these capabilities address rather specialized problems that do not have a large commercial market, although commercial technologies will provide many of the building blocks needed to realize the capabilities. Disasters are low-frequency events outside the normal planning horizons of most organizations, whose structure, operations, and IT systems are designed to ensure day-to-day efficiency rather than the resilience and scalability that disasters demand. As a result, current research and development efforts may not necessarily focus on developing IT capabilities in a manner optimized for disaster management.

As detailed in Chapter 4, IT R&D needs and opportunities are evident across a spectrum from adoption (off-the-shelf technology is available today) to adaptation (the technology is on the horizon and ready for transfer to disaster management) to development (the technology is on the horizon and requires development for use in disaster management) to applied research (disaster-management-specific research is required) to basic research (fundamental research is needed to develop new capabilities).

In government mission areas such as defense and energy, a research infrastructure has been built over decades to ensure long-term, mission-driven scientific and engineering advances—an effort that has included a long history of investments in IT. The Department of Defense, for example, funds a mix of shorter- and longer-term R&D carried out through the Defense Advanced Research Projects Agency (DARPA) and the service laboratories as in-house, university-based, and contract research. These investments are aimed at building a variety of capabilities, such as the military's transition to a capability for network-centric warfare, that are also relevant to disaster management. To make the sort of IT-enabled progress in disaster management that is envisioned in this report, the disaster management community should also devote significant attention and investment to a long-term research program.

A number of agencies could play a role in developing and implementing such a program. The directly relevant mission of the Department of Homeland Security's (DHS's) Science and Technology Directorate is "to protect the homeland by providing Federal and local officials with state-of-the-art technology and other resources."[2] Other agencies have relevant capabilities in terms of IT and disaster-related research programs, modalities, constituencies, and existing connections with particular research communities, including the National Science Foundation (NSF), the National Institute of Standards and Technology (NIST), DARPA, the National Oceanic and Atmospheric Administration (NOAA), and the research laboratories of the armed services.

In a number of federal programs, multiple agencies work jointly to tackle broad problems. One possible model for such an interagency program is the Earthquake Hazard Reduction Program, in which NIST has a lead role and the U.S. Geological Survey (USGS), FEMA, and NSF are participants. Similarly, for disaster management, a lead agency (logically DHS) could provide a clear single point of responsibility, coordinate activities, report on progress, and so forth; the lead agency would not be responsible for all aspects of execution, which would fall to all of the participating agencies and their contractors and grantees.

R&D activities also need to be well coupled to the parts of DHS that are responsible for mitigation, preparedness, response, and recovery activities to ensure that requirements are grounded in operational needs and to ensure that solutions can be transferred into federal operations and the parts of DHS responsible for developing policy to ensure that technological and organizational questions are considered together.

---

[2]See http://www.dhs.gov/xabout/structure/editorial_0530.shtm.

> **BOX S.1**
> **Key IT-Enabled Capabilities**
>
> - *More robust, interoperable, and priority-sensitive communications.* Disaster management requires robust, priority-sensitive communications systems capable of supporting interoperation with other systems. Providing these requires communication networks that are more resilient to disruption than today's commercial networks, that can last longer without utility power, that can expand capacity to meet emergency needs, that can autonomously reconfigure themselves, that can handle the range of communication needs and environmental conditions that arise in disasters, that have well-defined points of interoperability, and that are able to distinguish between and properly prioritize communications.
> - *Better situational awareness and common operating picture.* Situational awareness is the ability for actors in a disaster—from national coordinators to emergency responders to the general public—to have information about an incident, to understand what that information means in the context of the situation and their goals, and to project patterns and trends. The common operating picture is a shared understanding of a situation by a group of people who need to act together to achieve common goals. The aim is to improve a person's ability to do his or her job more effectively.
> - *Improved decision support and resource tracking and allocation.* Whereas situation awareness provides decision makers with information relevant to their tasks and goals, decision support focuses on assisting them in formulating prospective actions—helping them understand and assess characteristics and consequences of alternative courses of action and follow-up on decisions, closing the feedback loop from decision to result.

**Recommendation 2: The federal government should leverage the capabilities of its agencies to carry out multidisciplinary research in pursuit of six key IT-enabled capabilities—more robust, interoperable, and priority-sensitive communications; better situational awareness and a common operating picture; improved decision support and resource tracking and allocation; greater organizational agility for disaster management; better engagement of the public; and enhanced infrastructure survivability and continuity of societal functions—and establish a coordinating mechanism for those research activities.**

### Roadmapping as a Tool to Engage Stakeholders and Inform R&D Investments

Until fairly recently, the technology choices facing most disaster management organizations were few, with much of the investment focused on building specialized communications systems in close partnership with

> • *Greater organizational agility for disaster management.* The use of IT has enabled and driven changes to organizational structures and processes (e.g., more distributed decision making). Agility is at a premium in disasters because no one type of organization or group of organizations is always best suited for the variety of problems that arise. Related issues with significant IT implications include building rapport among people who do not share a history of cooperation and more quickly integrating the operations of multiple organizations.
> • *Better engagement of the public* by (1) supplying information and (2) making use of information and resources that members of the public can supply. Although IT is used today to alert and inform the public before, during, and after a disaster, more use could be made of new communications modalities, and information could be better tailored and targeted to the needs of particular populations. More attention should be paid to the information and resources held by the public because members of the public collectively have a richer view of a disaster situation, may possess increasingly sophisticated technology to capture and communicate information, and are an important source of volunteers, supplies, and equipment. One important factor is how to engage the entire population, given the existence of groups with cultural and language differences and other special needs.
> • *Enhanced infrastructure survivability and continuity of societal functions.* Large disasters upset physical infrastructure, such as the electric grid, transportation, and health care—as well as IT systems. IT infrastructures themselves need to be more resilient; IT can also improve the survivability and speed the recovery of other infrastructure by providing better information about the status of systems and advance warning of impending failures. Finally, IT can facilitate the continuity of disrupted societal functions by providing new tools for reconnecting families, friends, organizations, and communities.

a small set of vendors. Today, the set of technologies to choose from is much wider, and many more choices are possible in any particular area. But because disaster management is a system-level problem, there are no IT "silver bullets." Dramatic improvements in one area of technology or process may have relatively little overall impact unless other interconnected pieces are modified to make use of such advances. Too much invested in radios and not enough in logistics might mean, for instance, that one can call for help but cannot get it.

A clear vision of end-user goals, a detailed understanding of the individual pieces of the problem and their interrelationships, a detailed understanding of the required technologies, and defined paths for progress would help greatly to inform investment decisions. These are among the elements of a roadmap—an agreed-on, coordinated vision that can help organizations to plan development and investment strategies that can bring technologies together at the right time. Roadmaps are used in a number of sectors to accomplish this sort of alignment and cooperation.

A number of stakeholders, including first responders, public safety and emergency management agencies, government officials, medical providers, volunteer organizations, infrastructure and transportation system owners, vendors, IT researchers, and disaster researchers, have important perspectives on how to build on existing organizations and technology where possible and how to drive the creation of new, cost-effective technologies and organizational structures where needed. However, an institutional home is needed to launch and sustain such activity.

**Recommendation 3: The federal government should develop and regularly update an IT R&D roadmap for disaster management with the involvement of a full range of stakeholders.**

## IMPROVING ACQUISITION AND ADOPTION OF INFORMATION TECHNOLOGY

Adoption of IT for disaster management is challenging for a number of reasons, including the following (which are discussed in more detail in Chapter 3):

- Disaster management organizations often lack the resources to acquire valuable capabilities.
- The development and deployment of many promising technologies are risky and costly given the limited opportunities presented by commercial markets for these technologies.
- In most agencies with disaster management responsibilities, there is no person or unit specifically charged with tracking IT, identifying promising technologies, integrating them into operations, and interacting with IT vendors to make sure needs are addressed.
- Decisions regarding IT tend to be made independently by local organizations that must work together in disasters.
- Disaster management is concerned with environments that are intrinsically uncertain and unstable.
- Important sources of funds are typically only available once a disaster has been declared and also must be spent in a short time window.

### Diversified Acquisition Strategy and Attention to Design Issues

Disaster management has traditionally relied heavily on specially built technology and on a traditional "waterfall" acquisition model in which a full set of specifications is developed and a vendor is selected to build a system in compliance with the specifications. In the commercial world and in sectors such as defense, there is growing acceptance of a richer, more diversified acquisition strategy that employs a mix of traditional (purpose-built) systems, adoption and adaptation of commercial off-the-shelf (COTS) technology and services, and use of open source software, open standards, and community-driven development approaches. Such a strategy for disaster management should draw on the strengths of the traditional vendor community yet also foster bottom-up development as a complement to traditional acquisition practices and more formal top-down development; tap the nation's technology base; and also encourage the "can-do" spirit of commercial developers, professionals, and volunteers involved in disaster management.

As mentioned above, COTS technology alone is unlikely to deliver all of the desired capabilities. But it is important in two ways—(1) adoption and adaptation of readily available technologies offer a path for building up disaster management capabilities in the short term, and (2) the interoperability often characteristic of COTS technologies makes them useful as building blocks for disaster management systems.

**Recommendation 4: Federal, state, and local agencies should embrace a diversified acquisition strategy that includes increased use of commercial information technology and greater use of open source software and open standards development as a complement to more traditional acquisition approaches.**

Reliance on turn-key systems has meant that disaster management organizations have paid less attention to the underlying design issues that ultimately affect the functionality of their IT systems. Often technologies have been acquired as stand-alone products with little consideration for how they integrate with other technologies already in use, even within the same agency. However, with networking becoming increasingly pervasive, careful attention should be given to how each particular IT system fits into the broader context of interconnected systems. Off-the-shelf technology such as desktop computers or network routers can provide basic building blocks, and some elements of design are also well established such as the Internet Protocol standard for packet networks, but a domain-specific architecture understood and owned by the organization is also needed. To accomplish this, organizations have to develop the necessary technical and technology management capabilities.

**Recommendation 5: Disaster management organizations should work closely with technology providers to define, shape, and integrate new technologies as a coherent part of their overall IT system.**

The committee believes that more can be done to embrace proven approaches for IT acquisition. Best practices for acquisition include an emphasis on iterative development, increased opportunities to test and evaluate technology in practice together with realistic concepts of operations, and design and evaluation processes that allow for strong coupling among practitioners, researchers, and industry.

The committee also identified four design principles (discussed in more detail in Chapter 3) that have particular importance for disaster management systems:

- Build emergency management systems for effective scaling from routine to disaster operation;
- Exploit redundancy and diversity to achieve resilience;
- Design systems with flexibility, composability, and interoperability as core guiding principles; and
- Distinguish between the user interface and the underlying technologies used to deliver a capability.

### Training and Practice Through Routine Use of IT

Unless experience is gained through routine use or regular training, the full benefits of investment in IT systems are unlikely to be realized. Training, drills, and exercises all play an important role in the introduction of new technologies into organizational practice. Moreover, it is through routine use that the competence and confidence required to successfully use a technological capability, especially in the high-stress situation of disasters, are best developed. However, training large numbers of people to deal with infrequent events poses logistical challenges and is also costly.

**Recommendation 6: In the design, acquisition, and operation of IT systems, disaster management organizations should emphasize the incorporation of disaster response capabilities into the systems that support routine operations.**

### Measurement and Assessment to Enhance Effectiveness

As the saying goes, one can only manage what one can measure. Because the resources available for disaster management are limited, decision making always involves tradeoffs. Weighing the benefits from particular IT investments against the returns on other sorts of investment is challenging. Although having measures of effectiveness is necessary to making such assessments, few applicable metrics are currently available.

Above all, acquisition of IT and associated organizational changes should be driven by a focus on improving the effectiveness of those whose actions are integral to effective disaster management. The emphasis should be on measuring the resulting net effectiveness of disaster management activities, not the performance of the IT per se. For example, rather than focusing on the performance of particular systems (e.g., a firefighter radio system has 90 percent coverage), it is important to try to gauge net effectiveness (e.g., better communications coverage allows firefighters to better coordinate their response, improving capabilities for fighting forest fires by 20 percent). The development of such metrics is an area for further research.

**Recommendation 7: Disaster management organizations should employ metrics to inform cost-benefit decisions for investment in IT for disaster management and should make enhanced end-user performance a primary objective of disaster management acquisition programs.**

Independent mechanisms for assessment, such as the Transportation Safety Board, the U.S. Chemical Safety and Hazard Investigation Board, and the NASA Aviation Safety Reporting System, have proven useful in their respective domains. A critical requirement of these organizations is their independence from the agencies that have operational responsibilities. It may be advantageous to employ several organizations rather than a single national one, with each one focusing on a particular type of disaster or the range of disasters typical in a particular region. It is also important for assessments to be founded on multiple areas of expertise, including technical, social, and organizational dimensions. Obviously, the effectiveness of IT use is just one facet of an assessment of the overall effectiveness of disaster management activities for any particular incident, albeit an important one.

**Recommendation 8: Disaster management organizations should make use of independent mechanisms for assessing the effectiveness of disaster management operations, including the use of IT, and for disseminating lessons learned and best practices.**

### Systematic Collection of Data

It is well understood as a result of endeavors in many areas (e.g., aircraft accident investigations) that making significant improvements depends on putting in place processes that allow learning from experience. IT can play a critical role in enhancing the science of disaster management by helping to support continuous improvement. Collecting adequate and trustworthy data is essential for the feedback necessary to drive improvements in disaster management.

Doing so requires better documentation of disasters and the responses to them, including the role of supporting technology. The widespread introduction of IT into disaster management provides opportunities for large-scale, automated, comprehensive collection of data about operations, decision making, and situational knowledge throughout a disaster. This information can be processed after the fact to improve understanding of the process of disaster management. IT can also help to make these lessons learned available in real time, putting them into the hands of decision makers when they are most needed.

Accomplishing this goal requires enhanced technical capabilities to capture data during the course of a disaster. In some cases, it may be necessary to find ways to anonymize data so that accurate statistics can be gathered without the difficulty of dealing with potential liability issues. Policy changes (analogous to the adoption of Good Samaritan laws[3]) may also be needed to ensure that individuals involved in a disaster response can be protected from liability (e.g., amendments to the Health Insurance Portability and Accountability Act [HIPAA] may be needed). Legislation may also be needed to ensure that intellectual property, privacy, liability, and other concerns of information providers are addressed if they are required to share such data for research purposes.

## COUPLING RESEARCH AND PRACTICE

Effective development, use, and deployment of IT depend on a solid understanding of context and user needs. Moreover, the introduction of new IT often presents opportunities for new organizational approaches, and these opportunities should be considered in reorganization efforts. Similarly, successful technology development requires consideration of organizational context.

This issue of co-development of technology and organizational practice seems especially important at present. In the wake of Hurricane Katrina, a number of organizational structures, policies, and procedures are being examined. Also, a number of relevant technologies have reached a sufficient level of maturity to allow innovative organizational approaches to disaster management.

The interdependence of technology and practice suggests that developing a cadre of people with expertise in both disaster management and

---

[3]Good Samaritan laws shield people from liability in emergency circumstances provided that they have acted in a reasonable manner.

IT is likely to yield significant payoffs. Such a cadre of people is likely to be more astute at translating user requirements to technical needs. One basic need, of course, is to train a group of first responders and others responsible for disaster management in IT skills that go beyond those of a general user and to train a group of IT workers (e.g., database and system administrators and application builders) to have domain expertise in disaster management. Deeper expertise spanning both domains could be fostered through a number of mechanisms, including the following:

- A combination of fellowships, shorter-term visits to research centers, and other training and educational activities that help technology experts and other practitioners to stay abreast of the latest developments in both practice and technology;
- Field tests and field work conducted by IT researchers working with disaster management practitioners; and
- Combined disaster management-IT expert teams that jointly analyze the performance of processes and systems after a disaster.

**Recommendation 9: Disaster management organizations should support the development of a cadre of people with expertise in both disaster management and IT.**

Especially in light of the significant non-technical factors affecting adoption of IT for disaster management, it is critical to establish mechanisms that ensure that researchers are exposed to real problems and that practitioners are exposed to new technology opportunities. Because most practitioners are distributed across local agencies, forging such ties is likely to be harder in disaster management than in sectors like defense, but it is no less important.

Collaborative research centers could bring together experts from diverse domains in a neutral environment conducive to collaboration. Such centers could (1) develop a shared understanding of the challenges in all phases of disaster management from both a technological and an organizational perspective, (2) evaluate the application of technology advances to disaster management practice, (3) develop a culture and processes for transitioning knowledge and technology to the operational communities on a sustained basis, (4) build human capital at the intersection of IT and disaster management, (5) serve as repositories for data and for lessons learned from past disasters and disaster management efforts, and (6) provide forward-looking analysis to inform the development of technology capabilities, associated organizational processes, and roadmap development. A number of academic centers exist that offer a capacity for at least some of these efforts.

Practitioners in multiple disciplines could contribute to such centers, including scientists, engineers, and hazard and disaster researchers, and centers should include partnerships with federal, state, and local disaster management agencies. Indeed, it is critical that experienced and capable officials and operational elements of disaster management organizations be deeply involved in the work of these centers. One approach for engaging these groups would be to provide incremental funds to agencies specifically for working with researchers and developing next-generation technologies. Having multiple centers (rather than a single entity) also helps to ensure healthy intellectual competition, cross-fertilization of ideas, specialization in specific types of disasters and specific technology capabilities, and attention to the comprehensive needs of particular geographical areas.

Research centers could also act as a resource for agencies seeking to implement a diversified acquisition strategy and incorporate the latest best practices and a mechanism for disaster managers and responders to share experiences and communicate requirements to guide further technology developments.

**Recommendation 10: The federal government should sustain (and develop as needed) a network of research centers where IT researchers, hazard and disaster researchers, and disaster management practitioners can collaborate to study and evaluate the use of IT for disaster management from both a technological and an organizational perspective, to transition knowledge and technology to those who practice disaster management, to build human capital at the intersection of IT and disaster management, and to develop future IT capabilities.**

# 1

# Introduction and Context

This volume is the final report from the National Research Council's Committee on Using Information Technology to Enhance Disaster Management, which was established in response to a congressional request for a study to examine the use of information technology "to enhance crisis preparedness, response, and consequence management of natural and manmade disasters" (see Box P.1 in the Preface).

Drawing on a June 2005 workshop (see the agenda in Appendix C) and a series of briefings and site visits (listed in Appendix D), as well as the experience and expertise represented on the committee itself (outlined in Appendix E), the committee sought to identify promising applications of information and communication technology (hereafter referred to as IT) to disaster management, promising areas of research for improving the effectiveness of IT, and mechanisms that would enhance research, development, and deployment efforts. The resulting report is intended to inform federal, state, and local policy makers and public safety and emergency management professionals about future opportunities for the application of IT to disaster management. It is not intended as a comprehensive look at the complex, highly multidisciplinary topic of disaster management. Nor do the committee's findings and recommendations explicitly address tradeoffs between investments in information technology and other capabilities for disaster management or offer advice about levels of funding for IT or other disaster management activities.

This chapter provides a brief overview of challenges confronted in disaster management, focusing particularly on the use and role of IT;

presents several different ways of thinking about information and communication needs in disasters, which together provide a framework for understanding the various roles that IT plays in disaster management; and places the issue of IT use into the broader social context of disasters and disaster management.

## DISASTERS, DISASTER MANAGEMENT, AND INFORMATION TECHNOLOGY

Disasters are events that disrupt the normal functioning of the economy and society on a large scale (for more on terminology, see Box 1.1). Natural, technological, and willful (terrorist initiated) sources of disasters all cause dramatic losses of life and property.

---

**BOX 1.1**
**Terminology Used in Disaster Management**

A variety of terms are used in the fields of emergency and disaster management. Over time, a fairly standard set of definitions has emerged, as reflected in a series of reports from the National Research Council and other groups. Emergencies, disasters, and catastrophes, for example, are distinct events with important differentiating characteristics.[1] This report does not specifically consider "emergencies"—a term that connotes "everyday" events that can be handled within the normal operational limits of public safety agencies—nor does it distinguish between disasters and larger-scale events that might be called catastrophes, even though it is likely that the value of IT capabilities increases as the complexity and scale of communication problems become greater. Throughout this report, the term "disaster" can be read as "disaster and catastrophe."

This report uses the following set of definitions, adapted in part from *Facing Hazards and Disasters: Understanding Human Dimensions*:[2]

• *Disasters* are non-routine events in societies, regions, or communities that involve conjunctions of physical conditions with social definitions of human harm and social disruption. The term "disaster" has significant policy implications; for example, a declaration of an event as a disaster is needed before certain resources are made available.

• *Hazards* are a source of potential or actual harm. Hazards may be natural, technological, or willful in origin. Examples of natural hazards include floods, hurricanes, earthquakes, tsunamis, tornados, and so on. Technological hazards include industrial accidents and other human-made sources of potential harm. Bhopal and Chernobyl are examples. Terrorist attacks such as those on September 11, 2001, and the bombing in Oklahoma City are examples of willful hazards.

• *Incident* (or *event*) is the specific occurrence of a disaster. A single disaster incident may lead to additional incidents. For instance, an earthquake may lead to a tsunami and the tsunami may lead further to flooding. The term "incident" also

## BOX 1.1 Continued

has important bureaucratic meaning (e.g., incident period) that determines, for instance, who qualifies for financial assistance.

- *Risk* is a function of the likelihood (i.e., probability) of a specific event occurring and the potential consequences of harm should it in fact occur.
- *Hazard vulnerability* is the potential for physical harm and social disruption to societies and their larger subsystems associated with hazards and disasters. There are two general types of vulnerability: physical vulnerability and social vulnerability. *Physical vulnerability* represents threats to physical structures and infrastructures, the natural environment, and related economic losses. *Social vulnerability* represents threats to the well-being of human populations and related economic losses.
- *Hazard mitigation* is an ongoing effort to reduce the physical and social impact of future disasters. It includes interventions made in advance of disasters to prevent or reduce the impact. There are two major types of hazard mitigation: *Structural mitigation* involves designing, constructing, maintaining, and renovating physical structures and infrastructures to resist the physical forces of disaster impacts. *Nonstructural mitigation* involves efforts to decrease the exposure of human populations, physical structures, and infrastructures to hazardous conditions.
- *Disaster preparedness* includes actions taken in advance of disasters to deal with anticipated problems of disaster response and recovery. Actions include training and exercises to improve readiness; development and refinement of response and recovery plans; development, deployment, testing, and maintenance of systems used for disaster management; and public education and information programs for individuals, households, firms, and public agencies.
- *Disaster response* provides for the immediate protection of life and property, reestablishing control and minimizing the effects of a disaster. It encompasses the issuance and dissemination of predictions and warnings; planning and preparation immediately before an event (such as preparations following a hurricane warning); evacuation and other forms of protective action; mobilization and organization of emergency personnel, volunteers, and material resources; search and rescue; care of casualties and survivors; damage and needs assessment; damage control and restoration of public services; and maintenance of the political and legal system.
- *Disaster recovery* encompasses both short-term activity intended to return vital physical and social systems to operation and longer-term activities designed to restore these systems to their pre-disaster state. The concept of recovery encompasses both objective measures, such as reconstruction and assistance efforts, and the subjective experiences of disaster victims and processes of psychological and social recovery.

---

[1] E.L. Quarantelli, "Emergencies, Disasters and Catastrophes Are Different Phenomena," Disaster Research Center, University of Delaware, 2000; available at http://dspace.udel.edu:8080/dspace/handle/19716/674.

[2] National Research Council, *Facing Hazards and Disasters: Understanding Human Dimensions*, The National Academies Press, Washington, D.C., 2006, pp. 13-21. The report provides a detailed discussion of the importance of agreeing on basic definitions and the difficulty in doing so. See also D. Alexander, "The Study of Natural Disaster, 1977-1997: Some Reflections on a Changing Field of Knowledge," *Disasters* 21(4):284-304, 1997.

One of the essential characteristics of disasters is their complexity. Although disasters may have relatively discrete origins, their effects propagate and interact in ways that intensify the complexities and uncertainties of dealing with them effectively. One major result is that disasters must be responded to in an environment that can be overwhelming, unfamiliar, and disorienting. These challenges are quite familiar to experienced emergency managers and first responders, as manifest in a homespun sign found in many U.S. emergency operations centers (Figure 1.1a). This sign stands in marked contrast to a sign described by a reviewer of this report in draft form that lists what emergency managers aspire to—and often achieve despite the many obstacles—in a disaster (Figure 1.1b).

Disaster management is a multifaceted process aimed at minimizing the social and physical impact of these large-scale events. The difficult nature of disaster management is well illustrated by the Catastrophic Incident Annex to the National Response Plan, which lists some of the potential problems faced in the aftermath of a disaster (Box 1.2). Disaster

---

*The Emergency Operations Center:*
Where Uncomfortable Officials
Meet in Unfamiliar Surroundings
To Play Unaccustomed Roles
Making Unpopular Decisions
Based on Inadequate Information
And in Much Too Little Time                                      a


*Standing Orders for All Disasters:*
1. Establish/reestablish communication with affected area.
2. Secure and complete search and rescue.
3. Meet basic human needs for medical treatment, water, food, shelter, and emergency fuels; ice is a distant fifth (unless it's really hot).
4. Restore critical infrastructure.
5. Open schools and local businesses.
6. Begin the recovery.                                            b

---

FIGURE 1.1 Wording on two signs displayed in emergency operations centers illustrating (a) challenges to decision making and (b) basic goals. SOURCES: (a) Art Botterell, Office of the Sheriff, Contra Costa County, California, personal communication; (b) W. Craig Fugate, State of Florida Office of Emergency Management, personal communication.

# INTRODUCTION AND CONTEXT 19

---

**BOX 1.2**
**Catastrophic Incident Annex to the National Response Plan**

- The response capabilities and resources of the local jurisdiction (to include mutual aid from surrounding jurisdictions and response support from the State) may be insufficient and quickly overwhelmed. Local emergency personnel who normally respond to incidents may be among those affected and unable to perform their duties.
- A detailed and credible common operating picture may not be achievable for 24 to 48 hours (or longer). As a result, response activities must begin without the benefit of a detailed or complete situation and critical needs assessment.
- Federal support must be provided in a timely manner to save lives, prevent human suffering, and mitigate severe damage. This may require mobilizing and deploying assets before they are requested via normal NRP protocols.
- Large numbers of people may be left temporarily or permanently homeless and may require prolonged temporary housing.
- A catastrophic incident may produce environmental impact that severely challenges the ability and capacity of governments and communities to achieve a timely recovery.
- A catastrophic incident has unique dimensions/characteristics requiring that response plans/strategies be flexible enough to effectively address emerging needs and requirements.
- A catastrophic incident results in large numbers of casualties and/or displaced persons, possibly in the tens of thousands.
- A catastrophic incident may occur with little or no warning. Some incidents, such as rapid disease outbreaks, may be well underway before detection.
- The incident may cause significant disruption of the areas of critical infrastructure, such as energy, transportation, telecommunications, and public health and medical systems.
- Large-scale evacuations, organized or self directed, may occur. The health-related implications of an incident aggravate attempts to implement a coordinated evacuation management strategy.

---

SOURCE: Extracted from Department of Homeland Security, National Response Plan (amended May 25, 2006). Links to the National Response Plan and annexes are available at www.dhs.gov/nrp.

---

management is typically thought of as encompassing four phases: mitigation, preparedness, response, and recovery.[1]

Reducing the exposure to an event prior to its occurrence may be

---

[1] See, for example, Board on Natural Disasters, National Research Council, "Mitigation Emerges as a Major Strategy for Reducing Losses Caused by Natural Disasters," *Science* 284(5422):1943-1947, June 1999; U.S. National Committee for the Decade for Natural Disaster Reduction, National Research Council, *A Safer Future: Reducing the Impacts of Natural Disasters*, National Academy Press, Washington, D.C., 1991.

achieved by mitigation efforts aimed at preventing or reducing the threat and by preparedness measures meant to increase the capability or capacity of response and recovery efforts from anticipated problems in advance of an actual disaster event. Examples of mitigation include constructing buildings to accommodate impacts, identifying and measuring hazards to avoid putting social or physical assets in harm's way, and designing computer networks to degrade gracefully and recover from cyberattacks. Examples of preparation include detailed response planning, positioning resources prior to the onset of an event, setting up operations centers, training responders, and creating emergency management plans. Immediate response seeks to contain the event and minimize loss of life and injuries (rescue), health impacts, and property loss. Examples of immediate response include search and rescue operations. Sustained response seeks to restore critical systems to functionality and meet basic social needs. Examples of sustained response include restoration of sewers, water, and communications. Recovery seeks to minimize cascading impacts and facilitate long-term restoration to the pre-event situation.

Widespread efforts at managing disasters in a comprehensive fashion are a relatively modern phenomenon. Disaster management in the United States has historically been and remains a highly localized task that depends largely on local resources. Still, regional, state, and national efforts have grown out of the need to meet the increasing scale of disasters and the associated costs of managing them. (Box 1.3 identifies major milestones in the evolution of federal disaster management.)

Much progress has been made over the years in reducing the loss of life. Even loss of property from disasters is less than it might otherwise have been where various mitigation and preparedness strategies have been adopted and aggressive response and recovery efforts undertaken.[2] Yet, losses continue to increase.[3] Many factors have contributed to growing losses despite considerable progress in our understanding of them and in the practice of disaster management.

IT has been a major contributor to the progress that has been made.[4] Indeed, some of these applications have become so commonplace that it is easy to forget the improvements made over recent decades. One familiar

---

[2]See, for example, Board on Natural Disasters, National Research Council, "Mitigation Emerges as a Major Strategy for Reducing Losses Caused by Natural Disasters," *Science* 284(5422):1943-1947, June 1999.

[3]S.L. Cutter and C. Emrich, "Are Natural Hazards and Disaster Losses in the U.S. Increasing?" *EOS, Transactions, American Geophysical Union* 86(41):381-386, October 2005. As the title implies, this article focuses on natural disasters.

[4]Board on Natural Disasters, National Research Council, *Reducing Disaster Losses Through Better Information*, National Academy Press, Washington, D.C., 1999.

## BOX 1.3
## Major Milestones in the Evolution of the Functions and Profession of Federal-Level Disaster Management in the United States

- Development of the comprehensive emergency management taxonomy based on an all-hazards approach and the four phases of mitigation, preparedness, response, and recovery by the National Governors' Association in the 1970s.
- Establishment of the Federal Emergency Management Agency (FEMA) in 1979, which consolidated federal mitigation, preparedness, and response activities into one agency, reporting directly to the President. Responsibility for response to terrorist events, oil and hazardous materials releases, nuclear incidents, and health emergencies remained the province of other agencies, including the Department of Justice (DOJ), Environmental Protection Agency (EPA), U.S. Coast Guard, Department of Energy (DOE), and Department of Health and Human Services (DHHS).
- Publication of the Federal Response Plan (FRP) in 1991, providing a mechanism for organizing and coordinating the resources of 23 (later 27) federal agencies and departments and the American Red Cross.
- Amendment of the FRP in 1999 to include a terrorism annex for coordinating emergency management (termed consequence management) and law enforcement (termed crisis management) during a terrorist attack.
- Creation of the Department of Homeland Security (DHS) in 2002 and the movement of FEMA, the Office for Domestic Preparedness, and the U.S. Coast Guard under DHS, consolidating all emergency management functions into one department.
- Issuance of Presidential Decision Directive Number Five (HSPD-5) in 2003, directing DHS to create a national system for the management of all domestic incidents.
- Publication of the National Response Plan (NRP) in 2004, establishing protocols for the management of all incidents under DHS/FEMA direction. The NRP superseded and incorporated the FRP, the National Contingency Plan, and the Federal Radiological Emergency Response Plan.
- Issuance of the National Incident Management System (NIMS) in 2004, the creation of the NIMS Integration Center, and establishment of the requirement that state, local, and non-governmental emergency management organizations must be NIMS compliant in order to receive federal funds.

SOURCE: William L. Waugh, *Living with Hazards, Dealing with Disasters: An Introduction to Emergency Management*, M.E. Sharpe, Armonk, N.Y., 2000, p. 230; George D. Haddow and Jane A. Bullock, *Introduction to Emergency Management*, Butterworth Heinemann, Elsevier Science, Burlington, Mass., 2003, p. 275; Claire Rubin, *Disaster Time Line* and *Terrorism Time Line*, 2004, available at http://www.disaster-timeline.com.

example is application of IT to weather forecasting that has resulted in more accurate and timely warnings of hurricanes and floods.[5] IT has the potential for even greater impact on enhancing disaster management practice across all of its phases—mitigation, preparedness, response, and recovery—provided it is used consistent with the knowledge of hazards, disasters, and disaster management practices that has been gained from the diverse range of disciplines that contribute to that knowledge base.[6] Box 1.4 provides a sampling of uses and examples of particular technologies that illustrate the myriad ways in which IT is an integral part of disaster management today.

Responding to disasters involves such information- and communication-intensive activities as marshaling available resources and materiel, mobilizing and organizing sufficient skilled personnel, deploying them with those resources to where they are needed, and finally coordinating their actions. Specific tasks include establishing connectivity with potential resource providers, authorizing the use of resources and coordinating their use into something akin to a supply chain, integrating information from diverse (including ad hoc) sources, reducing the volume of data to relevant information for recipients, directing ongoing operations based on an overall awareness of the situation, adjusting and altering prior plans and commitments based on the evolving situation, and supporting collaboration and distributed decision making. The mitigation process is similarly complex and can involve many situation- and location-specific details, and it relies heavily on tools such as predictive models of the impacts of particular disasters. It is thus not surprising that IT has become a critical tool for facilitating the communications and information-processing activities in managing disasters.

The larger human and organizational context of disaster management was the subject of a recent National Research Council study. *Facing Hazards and Disasters: Understanding Human Dimensions* describes research undertaken during the past three decades by social scientists on hazards and disasters and recommends a continuing research agenda.[7] The report observes that the management of disaster mitigation, preparedness, response, and recovery has been aided by improvements in information

---

[5]National Research Council, *The Atmospheric Sciences Entering the Twenty-First Century*, National Academy Press, Washington, D.C., 1998.

[6]U.S. National Committee for the Decade for Natural Disaster Reduction, National Research Council, *A Safer Future: Reducing the Impacts of Natural Disasters*, National Academy Press, Washington, D.C., 1991.

[7]National Research Council, *Facing Hazards and Disasters: Understanding Human Dimensions*, The National Academies Press, Washington, D.C., 2006.

technology but cautions that events such as Hurricane Katrina provide a vivid demonstration that technology alone does not guarantee an effective organizational and public response to disasters.[8]

Indeed, an important lesson from past disasters is that applying IT in a vacuum (i.e., without considering the broader organizational and social context) may not only be ineffective but detrimental by, among other things, creating the perception that technology will solve all problems.[9] Put another way, there is no IT "Band-Aid" that will by itself overcome underlying organizational problems or problems rooted in systemic process, procedural, and policy issues that have never been reconciled.[10] Nor can IT alone address societal decisions such as settlement and land use patterns, construction standards and practices, and issues of social justice and equity. All of these factors and many others may increase vulnerability to hazards of large segments of the population and property. IT does, however, provide useful capabilities for tackling many of these challenges.

This experience is consistent with what is understood about the role that IT has played in productivity and quality advances in other sectors, ranging from defense to banking and finance. The empirical evidence shows, for example, that IT is not simply a tool for automating existing processes and that its real impact is as an enabler of organizational changes.[11] It is the complementary investment in decentralized decision-making systems, training, and business processes along with technology that allows organizational efficiency improvements.

There are a number of barriers to the adoption and use of IT in disaster management, growing out of the unique character of the institutions responsible, the organizational structure of the community as a whole, their need to focus on day-to-day operational missions, and their need to actively cooperate only under the most trying circumstances. Limited budgets, lack of expertise and other resources, demographic differences

---

[8]Ibid., p. 68.

[9]E.L. Quarantelli, "Problematical Aspects of the Information/Communication Revolution for Disaster Planning and Research: Ten Non-Technical Issues and Questions," *Disaster Prevention and Management: An International Journal* 6(2):94-106, 1997.

[10]Sharon Dawes, Thomas Birkland, Giri Kumar Tayi, and Carrie A. Schneider, *Information, Technology, and Coordination: Lessons from the World Trade Center Response*, Center for Technology in Government, University at Albany, State University of New York, June 2004; available at http://www.ctg.albany.edu/publications/reports/wtc_lessons/wtc_lessons.pdf.

[11]Jason Dedrick, Vijay Gurbaxani, and Kenneth L. Kraemer, "Information Technology and Economic Performance: A Critical Review of the Empirical Evidence," *ACM Computing Surveys* 35(1):1-28, March 2003.

> **BOX 1.4**
> **Some Examples of Uses of Information Technology in Disaster Management**
>
> • *Remote Human-to-Human Communications*—Starting with the first use of radios in coordinating disaster responses, voice communication over radio has been the primary role for IT in managing disasters. Indeed, by 1912 radio was recognized as critical to disaster response with the enactment of a law, in response to the *Titanic* disaster on April 14 of that year, which required all ships to have radios with two operators and auxiliary power and licensed transmitters.[1]
>
> • *Remote Sensing*—Networks of sensors are used in many ways, such as providing data for weather prediction and earthquake detection, to mitigate the impact of and to prepare for many natural disasters. Sensors are also extensively used to prevent or control human-made disasters. Doppler radar is used to identify and track hurricanes, tornados, and other weather phenomena. Networks of earth and structure motion detectors provide information about the severity and nature of earthquakes. Satellite imagery is used to map and plan operations in major wildland fires. Tsunami detectors provide advanced warning of the location and nature of tsunamis. Interferometric synthetic aperture radar (IfSAR), which uses an aircraft-mounted sensor to measure surface elevation, produces topographic imagery. Light detection and ranging (LIDAR) technology can measure the speed, distance, rotation, and chemical composition of a remote target, where the target can be either a clearly defined object, such as a building, or a diffuse object, such as a cloud. Other optical methods can also be used to remotely measure chemical composition of air masses to track toxic materials.
>
> • *Warning and Alerting*—Sirens systems have been used for over 100 years as a means to alert as many people as possible as quickly as possible.[2] One familiar IT-based warning system is the Emergency Alert System (EAS) established by the Federal Communications Commission (FCC) in November 1994; EAS replaced the Emergency Broadcast System (EBS) as a tool to warn the public

(e.g., urban versus rural), and the press of routine responsibilities also represent major constraints.

However, as was illustrated in testimony to the committee and in after-action reports of disaster responses, inventiveness, improvisation, and ingenuity have partially compensated for some of these shortcomings. Indeed, the often tacit knowledge of practitioners of disaster management about the realities of what works must also be incorporated if strategies for improving the use of IT in disaster management are to have maximum effect. The committee was mindful of this "can-do" spirit as it examined the needs and opportunities for using IT and ways of overcoming obstacles to its successful deployment and use.

about emergency situations. EAS includes AM and FM radio and broadcast and cable television (satellite operators can participate voluntarily). The National Oceanic and Atmospheric Administration (NOAA) Weather Radio All Hazards network of radio stations broadcasts continuous weather information directly from a nearby National Weather Service office along with localized warnings, watches, forecasts, and other hazard information. Recent years have seen efforts to extend warning systems to newer IT such as cell-phone- and Internet-delivered text messages.

- *Emergency Call Systems*—The 911 emergency call systems for wired phones have been deployed over a 20-year period across the United States and have now reached near universal coverage. The 911 service for mobile phones has recently been enhanced to provide location information about the caller.
- *Reporting*—Satellite communication hubs have enabled media to report from disasters. The public is also able to participate in new ways through the use of mobile phones, text messaging, and the Internet. While the technical capability for public participation in disaster reporting continues to grow, it has gone largely unrealized, even though recent disasters have shown that these alternative communication techniques are more robust than previously thought.
- *Modeling and Simulation*—Increasingly sophisticated models are being created of weather, storm surge, earthquake ground motion and shake intensity, toxic plume modeling, hazard prediction, and loss analysis. Culturally dependent models of population response would also be important for managing evacuations and other aspects of disaster management.

---

[1]University of San Diego, "History of Radio." See http://history.sandiego.edu/gen/recording/radio.html.

[2]Laura Olson, *Public Safety Best Practices: Talking Siren Technology: An Evaluation of U.S. Implementation of Early Warning Systems*, Metropolitan Washington Council of Governments, Washington, D.C., July 7, 2005. The report includes an examination of local government best-practice case studies of siren use across the United States.

## THE DIMENSIONS OF INFORMATION AND COMMUNICATIONS NEEDS IN DISASTER MANAGEMENT

Information and communications needs for disaster management are highly diverse in nature, reflecting the multiple purposes for information and communication and the different activities and information and communications requirements that occur at different times and locations with respect to a disaster.[12] Communications and information processing requirements in a disaster are very heterogeneous, varying according to context, use, time, latency, distance, and bandwidth. There are also many

---

[12]Board on Natural Disasters, National Research Council, *Reducing Disaster Losses Through Better Information*, National Academy Press, Washington, D.C., 1999.

types of information that can be communicated from many information sources. Indeed, the types of information available continue to grow with ongoing advances in IT. There is also a broad range of information actors and organizations involved in managing disasters; their ability to make appropriate decisions and function effectively can be greatly enhanced by IT and may depend on it for dealing with increasingly complex situations.

Given the heterogeneity of the information, the dynamics of the situation, and the diversity of actors, it is not surprising that there are a number of tensions that arise between more centralized, top-down, and planned disaster management activities and more decentralized, bottom-up, and ad hoc activities. For example, consider the following:

- The needs of "official" first responders versus those of emergent groups of people,
- Command-and-control decision making versus distributed decision making,
- The needs of first responders in the field versus the needs of higher-level decision makers, and
- The need for security and privacy protection versus the benefits of broadened access to information.

There are also inherent tensions between local governments and among federal, state, and local levels of government.

In each of these cases, the design and deployment of an IT system can make the tensions more acute and more visible. Moreover, IT cannot be used to paper over organizational problems—but its appropriate use may enable disaster managers to successfully accommodate a wider spectrum of disaster management activities, and do so more effectively and efficiently.

### Examples of the Kinds of Information Useful in Disaster Management

The first applications of IT to disasters were in the form of voice communications. Advances since then have led to many additional forms of information that have been included in disaster management practices to varying degrees, including text, geospatial data, video, sensor data, and collections of these and other types of data in databases or other electronic forms.

The number of available information sources has expanded considerably in recent years to include surveillance cameras; ground, air, and satellite sensors; telemetry from assets and personnel; unmanned vehicles; and eye witnesses with more technology. Some of these data sources have been well integrated into disaster planning. Other sources could improve situation awareness with efforts at better integration. The National Research Council study *Reducing Disaster Losses Through Better Information* catalogs a number of potential information sources (base data, scientific data, engineering data, economic data, environmental data, response data) and major types of information held and being gathered by federal agencies (e.g., base cartographic, land-use, seismic, hazardous site, demographic, aircraft route, river flow, and meteorological information).[13] Although some of these data sources are currently being used by disaster researchers for vulnerability assessment,[14] they are often inaccessible, unused, out of date, unusable, or inadequate for disaster managers, especially during response.[15]

Further advances in sensor technology (both pre-positioned and post-incident deployable) are likely to lead to opportunities for further improvements in both the volume and quality of data available. A number of factors affect data quality—completeness, timeliness, accuracy, and consistency—and advances should target all of them.

### IT Needs and the Incident Time Line

In thinking about the use of IT in a disaster it is useful to think of an incident time line consisting of three segments: *pre-incident*, *trans-incident*, and *post-incident*. In fact, the value of considering disasters and disaster management chronologically is unquestionable and taken for granted.[16] Disaster management can be viewed as roughly divided into three parts: (1) reducing exposure to and preparations for a hazard under routine, pre-incident circumstance; (2) preparations and actions immediately prior to and during an event; and (3) dealing with the consequences once it has occurred. Thinking in terms of time is also essential for understanding the different requirements for disaster management depending on the type of disaster. IT plays important roles in each time segment, and the committee considered the potential for increasing effectiveness in each one.

---

[13]Board on Natural Disasters, National Research Council, *Reducing Disaster Losses Through Better Information*, National Academy Press, Washington, D.C., 1999, pp. 13, 16-17.

[14]See, for instance, S.L. Cutter, B.J. Boruff, and W.L. Shirley, "Social Vulnerability to Environmental Hazards," *Social Science Quarterly* 84(2):242-261, June 2003.

[15]See, for example, Sharon Dawes, Thomas Birkland, Giri Kumar Tayi, and Carrie A. Schneider, *Information, Technology, and Coordination: Lessons from the World Trade Center Response*, Center for Technology in Government, University at Albany, State University of New York, June 2004; available at http://www.ctg.albany.edu/publications/reports/wtc_lessons/wtc_lessons.pdf.

[16]National Research Council, *Facing Hazards and Disasters: Understanding Human Dimensions*, The National Academies Press, Washington, D.C., 2006, p. 23.

## Examples of the Kinds of Data
## Communicated in the Response to a Disaster

One important kind of data communicated in a disaster is directives and authorizations for inter- and intraagency coordination. These are largely synchronous exchanges about where to go, where to meet, and reporting status. They can generally be accomplished through low-bandwidth mechanisms such as voice and text (e.g., e-mail, text messaging) and can be transmitted using both real-time media (voice or chat) and near-real-time media (such as e-mail). Coordination tasks result in interdependencies where Agency A cannot proceed with a task without authorization from or the arrival of Agency B—waits that can introduce significant delays in response activities if robust communications are not available.

Another important kind of data is requests for distributed decision making, especially logistics and planning. The content of these exchanges is akin to those that arise in dynamically creating a supply chain or a business enterprise where requests are tracked and processed. The exchanges may be more asynchronous where text and other files are sent and the time for a response is less immediate. Bandwidth requirements may generally be modest and traffic levels relatively low, but large files may need to be shared and databases kept synchronized.

Data are also needed to inform decision making at all levels and to help form a common operational picture. Relevant data include the following:

- *Human observations from direct response activities.* These are the reports from observers in the field, such as from local emergency operations centers, country transportation workers, pre-positioned trained observers, or other responders. The exchanges are largely one-directional, from the observer to a commander. Voice (e.g., cell and satellite phones) and text are both useful.
- *Geographic information system (GIS)-oriented data.* GIS information flows both to and from the field, with maps and projections such as flooding pushed to responders or tracking of personnel and assets collected by commanders. GIS information is typically in the form of high-resolution maps. If appropriate map sets have not been pre-positioned, large amounts of data may need to be transmitted; otherwise communications will generally take the form of updates and overlays to base maps.
- *Visual-oriented data.* Overhead imagery, satellite photos (before and after), and pictures and video of the disaster from other sources such as the public or other responding agencies can be extremely valuable. These data are inherently high-bandwidth and often need to be shared among

agencies, even if a common operating picture (a shared understanding of a situation by a group of people who need to act together to achieve common goals) is not established. Consider, for example, that a high-resolution overhead image of a coastline would help urban search-and-rescue units prioritize searches and allow transportation and law enforcement officials to determine avenues of access for supplies, controlling access, and so on.

- *Logistical information.* Information about the location and status of resources provides an important part of the common operational picture. Relevant data include databases and schedules describing where resources are, what and who has been dispatched to which affected areas, and so on; and what resources are needed by whom, where, and when, and the specific capabilities and limitations of those resources.[17]
- *Sensor data.* Information about the status of built infrastructure and environmental factors provided from pre-deployed instrumentation and devices deployed post-incident.[18]

### Examples of Sources of Data for Response

The source of data may have implications for their use as well as whether and how they are transmitted. Improving the effectiveness of data sources may include improving the usefulness of and access to the data. Valuable data sources exist for all phases of disaster management. Some sources of data for response include the following:

- *Data being "pushed" from the field.* During response, some sources of valuable data include data mapping damage, locations of responders and other resources, information on the status of response activities, and sensor data. High-fidelity (and thus high-bandwidth) data are required for some applications. For example, pictures or video of damage to a bridge could be transmitted to off-site experts for structural assessment.
- *Data being "pulled" from and "pushed" into the field.* A few examples of data responders' requests or data sent to them include field reports,

---

[17]This need for more detailed information about the specific capabilities and limitations of those resources is discussed in Sharon S. Dawes, Thomas Birkland, Giri Kumar Tayi, and Carrie A. Schneider, *Information, Technology, and Coordination: Lessons from the World Trade Center Response*, Center for Technology in Government, University at Albany, State University of New York, 2004; available at http://www.ctg.albany.edu/publications/reports/wtc_lessons/wtc_lessons.pdf.

[18]In the response to the September 11, 2001, attacks on the World Trade Center, environmental information was also needed about the construction materials and possible hazardous materials in damaged or destroyed buildings.

imagery and map updates, and status information on resource deployments.

- *Discovery services.* The Internet offers the prospect of identifying and creating information resources dynamically from a wide variety of official and non-official sources.

### Tactical Versus Strategic Operations

Tactical operations focus on the response operations in the affected area. They may involve stabilizing the situation sufficiently to carry out those operations. (In the case of a terrorist incident, this could involve capture or neutralization of the threat as well as responding to it.) Decisions are often immediate, based on direct observation and a priori knowledge. Any available information useful for gaining a broader understanding of the situation to aid in the decision-making process may help answer questions critical to tactical operations such as, What resources are nearby that I can use? How extensive is this problem? Can neighboring resources/units be directed my way? When can I expect help to get here? A secondary role of tactical operations is as general information gatherers for the strategic operations.

Strategic operations, by contrast, are essentially "enterprise" decisions and may span weeks to months. Information flow for strategic operations is highly computer-centric and more akin to the flow of information in a supply chain management system.

A particular challenge—and one where IT can play an important role—in disasters is connecting tactical operations and strategic operations. The physical distance between the tactical and strategic decision makers, as well as the differing time spans for decisions, poses additional challenges for cooperative work and information sharing. Another area requiring further research is in understanding what the roles should be for strategic and tactical operations and how IT should be structured in order to properly support those roles.

## A FRAMEWORK FOR CONSIDERING USE OF INFORMATION TECHNOLOGY IN THE BROADER SOCIAL CONTEXT

As suggested above, both successes and failures in disaster management depend on the effectiveness and resilience of human as well as technological systems. As a result, broad statements about IT failures during disaster, although often true, can be a major source of confusion about the complex sources of disaster management failures. A four-layer "stack" model developed by the committee illustrates the range of issues that are sometimes lumped under the rubric of "communication," "interoperability," and "information technology" issues. The model includes the following elements:

1. *Organizational and social context* comprises the goals, metrics, priorities, and beliefs of each organization involved, as well as those of "meta-organizations" such as an incident command structure or an emergency operations center that involves multiple organizations. It is concerned with the purpose, content, and partners in communication. The social side goes beyond the logical layer that facilitates transmitting content. It includes the cultural and organizational constraints and workarounds to organizational barriers, such as informal social networks between trusted friends in different organizations. One source of "communication" problems at this level can be simple misalignment between the goals and priorities of different organizational elements; for example, at a potential terrorism site there might be a "disconnect" between organizations focused on criminal investigation and others concerned chiefly with rescue or restoration of services.

2. *Human behavioral context* includes the many variables of individual human performance, including skill sets, training, experience, health, personal stress, and other personal factors. Despite efforts at standardization, human beings inevitably bring a degree of variability into the execution of procedure and the pursuit of organizational goals and values. This is not necessarily a bad thing; indeed human originality and adaptability are often critical to meeting unforeseen challenges. But they can insert an only partly controlled variable into the performance of carefully planned processes and can give rise to problems that are sometimes mischaracterized as technology-related communications problems, especially under conditions of high stress or uncertainty typical in disasters.

3. *Procedural and policy framework* refers to predictable patterns of behavior. Sometimes these are formalized and documented, but often they are unwritten and even unconscious artifacts of an organizational or disciplinary culture. Procedures are often event-driven, that is, expressed in the form "when X happens, do Y." A great deal of implicit knowledge about the immediate system and its environment is encoded in such procedures. As a result, they can be confounded by the profound changes in context that may accompany a disaster. For example, an emergency communications plan can be disrupted by loss of, or interference with, expected technical channels. Without an effective way to devise and transition to an alternate plan, such disruptions can lead to perceived major failures of communication, even when significant technical capability remains.

4. *Technology* includes the bulk of what is frequently understood in the terms "communications," "interoperability," and "information technology." Technology is the medium through which communications and infrastructure needs are met, and can be thought of as the physical layer of communications. It includes all of the capital infrastructure investment for communications and information technology. Although technical failures are by no means uncommon, they can frequently be circumvented using alternate technologies, provided the procedures, skills, and organizational will remain to implement such expedients.

This conceptual model presents a number of useful insights. First, *problems* (and the perceptions of problems) *tend to propagate downward* (from 1 to 4) through the stack, so that various non-technical issues can end up being framed as technology failures. For example, police and firefighters at a traffic accident might have subtly different organizational priorities. The firefighters might be focused on the well-being of victims at the scene, while the police might be tasked with reestablishing unhampered traffic flow for the larger community. This organizational difference might lead to personal and procedural conflicts that ultimately might be (mistakenly) characterized as a "communication problem" and then (also mistakenly) interpreted as a failure of "interoperability," which is frequently assumed to be a technical issue.

Second, *change tends to propagate upward* (from 4 to 1) through the stack. Effective use of new technologies requires and enables new procedures, which in turn require new skills and create new challenges, to which organizations ultimately must adapt. For example, in many large organizations, computer-based word-processing software was first introduced in a "word-processing pool" office, by analogy to previous typing and dictation pools. Over time the opportunities that provided for faster and more flexible service moved the new technology out to secretarial desks in the operating departments, and eventually onto the desktops of commanders and executives. The word-processing pool, and in some cases the secretary as well, faded into organizational history.

Third, *many interoperability and data-sharing challenges are not fully or even mostly technical* in nature. Indeed, as noted in the report summarizing a workshop convened as part of this project, better "human organization, willingness to cooperate, and a willingness of government at higher levels to listen to those at local levels who really do the work and who are the actual responders are all critical factors in making better use of information technology for disaster management."[19] As a result, many inter-

---

[19]See National Research Council, *Summary of a Workshop on Using Information Technology to Enhance Disaster Management*, The National Academies Press, Washington, D.C., 2005, p. 2.

operability and data-sharing challenges may not be amenable to technical solutions alone—or at all.

## LEVERAGING INFORMATION TECHNOLOGY TO IMPROVE THE EFFECTIVENESS OF DISASTER MANAGEMENT

Chapter 2 builds on the discussion in this chapter to outline a vision of the potential for IT to improve the effectiveness of disaster management in all its phases. The vision encompasses six areas of IT-enabled capabilities identified by the committee as having particularly significant potential. Chapter 3 examines mechanisms for focusing IT research and development on disasters and disaster management in a way that reflects disaster research and the experience of practitioners. It also explores mechanisms for improving the transfer and adoption of IT into practice. Chapter 4 sketches a potential IT research agenda based on the vision elaborated in Chapter 2 and driven by the mechanisms described in Chapter 3.

## 2

# The Potential to Enhance Disaster Management: Key IT-Based Capabilities

How could better application of information technology (IT) to disaster management reduce the human and economic costs of catastrophic events? This chapter outlines a vision for IT-enhanced disaster management in terms of six areas of IT-based capabilities. Three scenarios developed by the committee (presented in Appendix A) describe specific fictional disasters to help place those capabilities in the context of existing IT use in disaster management and to highlight how progress would have tangible positive impacts.

Reducing the impact of disasters requires a complex mix of technical and social endeavors, and no single prescription or discipline can provide all the answers.[1] Indeed, disaster researchers have frequently expressed concerns that technology not be viewed as a panacea.[2] The committee shares the view that enhancing disaster management requires attention to

---

[1]See, for example, National Research Council, *A Safer Future: Reducing the Impacts of Natural Disasters*, National Academy Press, Washington, D.C., 1991.

[2]See, for example, E.L. Quarantelli, "Problematical Aspects of the Information/Communication Revolution for Disaster Planning and Research: Ten Non-Technical Issues and Questions," *Disaster Prevention and Management* 6(1), 1997. See also Sharon S. Dawes, Thomas Birkland, Giri Kumar Tayi, and Carrie A. Schneider, *Information, Technology, and Coordination: Lessons from the World Trade Center Response*, Center for Technology in Government, University at Albany, State University of New York, 2004; available at http://www.ctg.albany.edu/publications/reports/wtc_lessons/wtc_lessons.pdf.

technological, organizational, and social factors and depends on a solid understanding of disaster management as well as the technologies.[3]

Nonetheless, IT represents an important point of leverage for enhancing disaster management.[4] Briefings to the committee suggested that progress continues to be made toward ever more effective use of information technology to enhance disaster management. Better preparation and training of public safety officials and the public, improved mitigation and prevention measures, more efficient and effective response, and more rapid recovery are all possible.[5] Furthermore, the public has high expectations that technology that it sees deployed ever deeper into other societal systems will be applied to improve the handling of disasters.

This chapter discusses six key IT-based capabilities that were selected by the committee because they (1) have the potential to address major problem areas in current disaster management practice and (2) represent areas where there appears to be significant potential for further advancement of the current state of the art. These capabilities span areas that could improve hazard mitigation, disaster preparedness, disaster response, and disaster recovery. They also aim to address requirements of practitioners at all levels—first responders, local or regional emergency managers, and national emergency managers. Improving these capabilities is applicable to addressing natural, accidental, and terrorist-induced disasters, though some capabilities may be more specific to one type.

Table 2.1 lists these capabilities together with examples of near-term, mid-term, and long-term opportunities for technology development that were identified by the committee. Some of these technology areas are already the focus of significant federal research and development investment. For example, self-managing and repairing networks are the focus of a Defense Advanced Research Projects Agency (DARPA) research program. Some technologies are the focus of considerable research and development investment from the private sector, such as wireless mesh

---

[3] J.-L. Wybo and H. Lonka, "Emergency Management and the Information Society: How to Improve the Synergy?" *International Journal of Emergency Management* 1(2):183-190, 2003.

[4] Robin Stephenson and Peter S. Anderson, "Disasters and the Information Technology Revolution," *Disasters* 21(4):305-334, December 1997. This paper reviews the application of IT to disaster management from 1970 through the middle 1990s. It identifies a considerable body of literature showing the evolution of the use of IT in disaster management from its beginnings to becoming "an indispensable component of disaster operations worldwide."

[5] John Pine, "Research Needs to Support the Emergency Manager of the Future," *Journal of Homeland Security and Emergency Management* 1(1), Article 3, 2004. Originally presented at the National Academies' conference "The Emergency Manager of the Future: A Disasters Roundtable," June 13, 2003.

TABLE 2.1 Key IT-Based Capabilities for Disaster Management and Related Promising Technologies

| Key Capability | Promising Technologies — Near Term |
|---|---|
| More robust, interoperable, and priority-sensitive communications | Cellular<br>Wireless networking<br>Redundant and resilient infrastructure<br>Internet/IP-based networking |
| Improved situational awareness and a common operating picture | Radio-frequency identification for resource tracking and logistics |
| Improved decision support and resource tracking and allocation | Online resource directories<br>Commercial collaboration software and file sharing |
| Greater organizational agility for disaster management | Computer-mediated exercises<br>Portable unmanned aerial vehicles and robots |
| Better engagement of the public | Automated, multimodal public notification and resource contact systems<br>Multimodal public reporting capabilities<br>Validated online information sources<br>Reverse 911 capability (i.e., two-way emergency reporting) |
| Enhanced infrastructure survivability and continuity of societal functions | Mobile power generators<br>Redundant radio systems<br>Dynamic stockpiled supply management |

| Midterm | Long Term |
|---|---|
| Mobile cellular infrastructure<br>Intelligent spectrum sharing<br>Multiple input/multiple output wireless systems<br>Integrated voice and data<br>Policy-based access control mechanisms | Software-defined radios<br>Delay-tolerant networking<br>Passive and active embedded conductors and relays for enhanced communication in buildings, rubble, and underground<br>Policy-based routing and congestion management<br>Self-managing and repairing (autonomous and adaptive) networks |
| Embedded, networked sensors<br>Routine information fusion<br>Publish/subscribe information dissemination<br>User-centered situational awareness information presentation | Semantic routing<br>Data mining across diverse information sources<br>Calibrated information confidence tools<br>Deployable sensor networks<br>Automated information fusion from diverse sources<br>Network and information security<br>Augmented cognition |
| Dynamic responsibility charting<br>Intelligent adaptive planning tools | Decision sentinels<br>Distributed emergency operation centers<br>Resource use modeling<br>Coordinated transportation, communication, and decision support<br>Computer-assisted decision-making tools |
| Event-replay tools<br>Online repositories of lessons learned<br>Integrated ad hoc data-collection tools (blogs/wikis) | Continuous learning tools<br>Computer-assisted disaster simulation training<br>Distributed, scalable, survivable data logging<br>Dynamic capability profiling and credentialing<br>Collective sensemaking |
| Volunteer mobilization systems<br>Distributed, dynamic private resource directories<br>Enhanced two-way communication with public | Automated public reporting tools<br>Optimized data formatting for differing presentation devices |
| Network redundancy<br>Renewable power sources<br>Embedded sensors for nondestructive asset evaluation | Risk management tools with uncertainty modeling<br>Resilient "smart" materials and structures for infrastructure survivability |

networks and collaboration software. Different actions are likely to be required to move technologies toward deployment in the field depending on where they are in the technology pipeline (discussed in Chapter 3), the degree of specialized technological adaptation required for use in disaster management, and the need for organizational changes.

The sections that follow examine each of the six key capabilities in more detail.

## MORE ROBUST, INTEROPERABLE, AND PRIORITY-SENSITIVE COMMUNICATIONS

During a disaster, both commercial and public safety communication infrastructure—telephone lines, radio towers, communication switches, network operation centers, and the requisite power needed—is often degraded or damaged. Simultaneously the demand on communication increases from the public and from first responders. Mobile communication demand is especially acute. New users from external public safety jurisdictions enter the disaster region. The public may have to be mobilized for evacuations. Moreover, environments confronted by first responders, such as collapsed buildings, place unusual requirements on communications capabilities. And, finally, hostile action may compromise even those resources that survive the initial disaster.

Simple availability of communications is thus a critical starting point for disaster response. Communication robustness can be improved by applying well-understood techniques for improving the availability of systems, that is, hardening infrastructure, improving network resilience and adaptability, providing redundancy and diversity, improving component robustness, and optimizing recovery speed. Notably, many communications problems are caused by the destruction of communication devices or communication lines, by the loss of power to radio towers, switches, and other infrastructure, and by the lack of ability to recharge handsets and other mobile devices. These are often the result of damage to physical structures (buildings, cell towers). Although it is not economically feasible to harden all relevant equipment for worst-case scenarios, improvements are certainly possible.

Commercial services could provide a valuable complement to dedicated disaster management communications systems by providing redundant infrastructure for voice and data communications. Generally implemented on a different technology base than that of dedicated systems, they also add diversity. Commercial infrastructure, such as the cellular network, is also, by design, highly interoperable. (Mobile callers can readily communicate with a subscriber to their carrier, any other carrier, or a landline subscriber.) The participation of wireless Internet service providers in Hurricane Katrina response and relief efforts hints at the possibilities of embracing and integrating another type of commercial technology into disaster management practice. Of course, one reason for deploying separate public safety radio systems is that they are designed to be more resilient that commercial infrastructure. And public safety radios are designed to much more stringent operational standards than mobile phones. Yet, both commercial and public safety communications infrastructures have suffered breakdowns in recent disasters.

One critical issue that arises in considering commercial services is priority access. When using commercial services, priority could be provided in a variety of ways. For example, it is possible to set up a cellular network in a spectrum band that is dedicated to disaster management, though cellular handsets would have to be modified (at a significant additional cost) to support this (and the spectrum allocated). Alternatively, a commercial service can be used, with arrangements made (preferably prior to an incident) to reserve a certain fraction of the capacity for prioritized disaster management traffic. Of course, priority access to commercial services requires a contractual and regulatory framework as well as technology adaptation.

Commercial technology (and adaptations of commercial technology), whether used as part of commercial services or incorporated into dedicated disaster communications systems, could be used to more quickly leverage the latest technology advances and capabilities. Commercial technologies provide a range of capabilities supporting voice and data at rapidly increasing transfer rates. Two examples of commercial technologies widely embraced by the public which hold potential for disaster management are cellular telephony and 802.11 standards-based wireless networking. Cell phones are now used by a large majority of people on a daily basis, so it is natural to continue to use them in a disaster. Besides traditional voice calls, cell phones often also support push-to-talk capabilities, text messaging, and Web access, all of which could be usefully employed. Wireless networking is also becoming ubiquitous in many areas and is supported on laptops and handheld devices. More generally, Internet Protocol (IP)-based communications networks could allow support of emerging IP-based multimedia services, high-data-rate access, and mission-critical tactical group voice and interoperable communications during emergency responses.[6] All these technologies are widely used and relatively inexpensive. Their familiarity addresses another issue re-

---

[6]Krishna Balachandran, Kenneth C. Budka, Thomas P. Chu, Tewfik L. Doumi, and Joseph H. Kang, "Mobile Responder Communication Networks for Public Safety," *IEEE Communications Magazine* 44(1):56-64, 2006.

lated to communications robustness—the tendency to fall back on technology used routinely.[7]

Although some commercial communication technologies have been successfully adopted by some first responders, there is certainly additional scope for using these technologies, often with minimal adaptation, to improve disaster response. Rapid deployment of wireless, cellular, and satellite infrastructure to replace infrastructure damaged in a disaster is one such opportunity. Mobile, rapidly deployable infrastructure has the advantage of leveraging commercial technology without necessarily relying on commercial service providers. Mobile infrastructure can be brought in after an event to quickly reestablish communications, (partially) replacing infrastructure lost during the event. Mobile cellular technology might also be used to bring cellular infrastructure to rural or other areas lacking pre-deployed cellular service just as it has already been demonstrated to bring such service to areas where existing cellular infrastructure was destroyed.[8]

There are, however, important communications problems that arise in disaster management, especially in response activities, that are unlikely to be addressed in the commercial market. One notable example is meeting communications needs inside buildings or other enclosed spaces, including damaged structures. Possible approaches include mandating the deployment of hardened repeaters in buildings, developing and deploying special low-frequency radios, and developing and deploying low-cost "bread crumb" repeaters for first responders. Another example of such a communications problem is the congestion (both local and system-wide) that arises when and where communication is most needed, resulting in a spike in communications traffic, especially in the area where infrastructure is likely to be most damaged. Handling surging demand that coincides with reduced capability will require new approaches.

Not surprisingly, given the number of organizations that must come together to cope with a major disaster, the interoperability of communications and other IT systems is often cited as a major concern. The overall evolution of communications systems (and, indeed, IT more broadly) in disaster management has been characterized by local, agency-level acquisition and deployment driven by local budgets from local taxing bodies

---

[7]National Research Council, *Summary of a Workshop on Using Information Technology to Enhance Disaster Management*, The National Academies Press, Washington, D.C., 2005, p. 4.

[8]Presentation of Mark Koro, Qualcomm, to the committee on December 12, 2005, on Qualcomm's deployment of mobile cellular on wheels, including mobile switches on wheels, during Hurricane Katrina.

and by local priorities. This has led to the creation of a heterogeneous mixture of voice and (more recently) data systems across the United States. The result is that different public safety agencies (e.g., police, fire, and emergency medical), even in the same community, are often unable to communicate or share information with each other. Interoperation is not typically considered when IT is acquired. Thus, it is not surprising that limited technical interoperability exists.

Concerns about public safety communications interoperability are not new, though they have received increasing attention in recent years. For example, the Public Safety Wireless Advisory Committee (PSWAC), in a 1996 report to the Federal Communications Commission and the National Telecommunications and Information Administration (NTIA), concluded that "unless immediate measures are taken to alleviate spectrum shortfall and promote interoperability, public safety will not be able to adequately discharge their obligation to protect life and property in a safe, efficient, and cost-effective manner."[9] A 1997 National Institute of Justice (NIJ) study that surveyed state and local law enforcement agencies confirmed and quantified a number of issues identified in the PSWAC report.[10]

Additional emphasis has been placed on interoperability and associated issues in the wake of the 9/11 attacks, the debate on the transition to digital television, and the response to Katrina. The 9/11 Commission, for example, cited first responder voice communications interoperability as a considerable issue in the response to the attacks and concluded that the highest-priority remedy was assignment of additional radio-frequency spectrum as a way of achieving interoperability. It recommended that Congress enact legislation providing for the "expedited and increased assignment of radio spectrum for public safety purposes."[11] The 9/11 Public Discourse project report-card-like "Final Report on 9/11 Commission Recommendations" gave an "F" to progress on providing adequate radio spectrum for first responders, citing lack of progress freeing up the analog television broadcast spectrum and reserving some of it for public

---

[9]Public Safety and Wireless Advisory Committee (PSWAC), *Final Report of the Public Safety and Wireless Advisory Committee,* presented to the FCC and NTIA September 11, 1996, p. 2; available at http://ntiacsd.ntia.doc.gov/pubsafe/publications/PSWAC_AL.PDF.

[10]Mary J. Taylor, Robert C. Epper, and Thomas K. Tolman, "Wireless Communications and Interoperability Among State and Local Law Enforcement Agencies," NCJ 168945, National Institute of Justice, Washington, D.C., January 1998.

[11]National Commission on Terrorist Attacks Upon the United States (also known as the 9/11 Commission), *The 9/11 Commission Report: Final Report of the National Commission on Terrorist Attacks Upon the United States,* Government Printing Office, Washington, D.C., 2004, p. 397; available at http://www.9-11commission.gov.

safety purposes.[12] More recently, legislation was enacted that calls for a handover in 2009.[13]

First responder interoperability is often cited as a major problem in responding to disasters, and recommendations aimed at addressing interoperability frequently appear prominently in after-action reports on major disasters. Indeed, improved first responder communications would have important benefits in terms of enabling communications across jurisdictions and among fire, police, and medical services. The issue has deservedly received attention from the public, government officials, and lawmakers. However, to place this issue in context, it is worth noting that interoperability is only one of many significant communications issues facing first responders, and first responder communications are not the only technology-related disaster management need (see Box 2.1).

Furthermore, interoperability issues arrive in many guises, from compatibility of waveforms and message formats (technical), to terminology and definitions (semantic), to practices and procedures (organizational). Much of the public attention has been focused on voice communications, but within the public safety community, data communications interoperability is an increasingly critical component and central part of any communications system. Data communications interoperability, while having some issues in common with voice communications, raises a number of specific issues that arise when sharing information from different sources. The 1997 NIJ study noted the trend toward increasing reliance on information sharing and the importance of data communications interoperability to enable it. (For an overview of interoperability initiatives, see Appendix B.)

A number of efforts are underway to increase technical standardization, and a number of technical solutions have been developed that allow systems such as first responder radios to be "patched" together. However, interoperability should not be viewed as solely a technical problem. The harder problem is deciding when the various users across these interoperable systems should talk to each other, the protocol for doing so, who can make those decisions, and how teams get formed and dissolved.

Although information could, in principle, flow arbitrarily in distributed networks, in order to act, some sort of structure is needed. Most information is hierarchically organized, but there are many different possible hierarchies, reflecting the need and point of view of the creators.

---

[12]9/11 Public Discourse Project, "Final Report on 9/11 Commission Recommendations," December 5, 2005.

[13]Deficit Reduction Act of 2005, S. 1932, Sec. 3403, "Analog Spectrum Recovery: Hard Deadline," Congressional Budget Office, Washington, D.C., January 27, 2006.

## BOX 2.1
### Interoperability in Context

Interoperability is only one of many significant communications issues facing first responders. Consider the following, for example:

- *One cannot interoperate without being able to operate in the first place.* Simple availability of communications (i.e., operability) is an important consideration because disasters often cause damage to communications systems. Both agility, which makes it possible to more rapidly reconstitute communications, and robustness, which makes it easier to keep communication systems operating, are at a premium. Moreover, environments such as collapsed buildings that confront first responders place unusual requirements on communications capabilities.
- *No single communication or information system is likely to suffice in a major disaster, no matter how comprehensive its scope or how much investment has been made in its interoperability or technical robustness.* A variety of communications systems may end up being pressed into service, reflecting the value of redundancy.
- *Technical communications interoperability does not address the challenges of data interoperation among organizations.* Disaster response often requires access to information held by multiple independent organizations. However, each organization's information is structured first to benefit its own routine needs and secondarily to meet non-routine external needs. Experience from many domains shows that working across these so-called silos cannot be addressed simply by imposing a single standard across all organizations.[1]

First responder communications are far from the only technology-related disaster management need. Consider the following:

- The very first individuals on the scene of a disaster are likely not first responders but members of the public.
- Many other actors and organizations (other than first responders) are involved in managing disasters; their ability to make appropriate decisions and function effectively also depends on IT.
- Communications and information-processing requirements in a disaster are very heterogeneous, varying according to context, use, time, latency, distance, and bandwidth.

---

[1] Related to this, there may also be different regulatory conditions among organizations—for example, the Health Insurance Portability and Accountability Act (HIPAA) governs how patients' health care information can be handled by health care organizations. In the wake of Hurricane Katrina, the Department of Health and Human Services issued guidance to remind health care organizations about HIPAA's emergency provisions. See U.S. Department of Health and Human Services (HHS), "Hurricane Katrina Bulletin: HIPAA Privacy and Disclosures in Emergency Situations," HHS, Washington, D.C., 2005; available at http://www.hhs.gov/ocr/hipaa/KATRINAnHIPAA.pdf.

There are numerous examples where elements are covered in different hierarchies, and for good reasons. Efforts to determine a priori a "correct" information hierarchy are often ineffective. Thus, simply making communications and systems technically capable to interact may create more problems than it solves, unless the deeper meaning of interoperation is understood and addressed.

## IMPROVED SITUATIONAL AWARENESS AND A COMMON OPERATING PICTURE

Situational awareness capabilities, like communications capabilities, have received considerable attention recently, especially in the aftermath of Hurricane Katrina. Involving much more than "having all the information," situational awareness is an achieved mental state; it does not begin or end with the presentation of data on a display. It is the degree to which one's perception of the situation reflects reality. Improving situational awareness capabilities must include advancing and integrating technology that assists disaster managers in building an accurate and complete mental model, in addition to improving the amount and quality of the information available. Reflecting the difficulties of achieving situational awareness, one researcher has compiled a list of the "demons of situational awareness" (Box 2.2). Just as it is important to caution that solving technical interoperability is not enough to achieve interoperation among organizations, it is important to clearly caution that solving technical situational awareness is not enough. IT could enable implementation of solutions to situational awareness demons, but the solutions must come from an understanding of the human dimensions of those demons, and the IT systems incorporate that understanding.

As described in Chapter 1, an increasing amount of information can potentially be brought to bear in disaster management. More information about a disaster may initially seem like a good thing. Yet, data from disparate sources can be difficult to assimilate into useful information because of a multitude of formats, the difficulty of placing sensors where they are needed, and the difficulty of communicating sensor data to those who need it. Moreover, without filtering by human or automated mediators, those receiving the information are likely to become overloaded and to ignore excessive inputs as distractions or to devote already-scarce resources to monitoring or processing them.

Research to identify leverage points for IT to augment and amplify the human ability to make sense of data may improve the effectiveness of disaster management. "Sensemaking" is "the process of searching for a representation and encoding data in that representation to answer task-

## BOX 2.2
### Eight Major Demons of Situation Awareness

• *Attentional tunneling.* Good situation awareness is highly dependent on the ability to monitor multiple sources of information simultaneously. Unfortunately, people frequently lock in on certain aspects of the situation they are trying to process—get tunnel vision—and will either intentionally or inadvertently ignore all other aspects. This leads to their situation awareness of those other aspects becoming quickly outdated. People may believe this is the proper course because the aspect that they are focusing on appears to them to be the most critical. However, keeping at least a high-level understanding of what is happening across the board is a prerequisite to being able to know that certain factors are indeed still more important than others. It is often the neglected aspects of a situation that prove to be fatal. Human-information interaction systems that foster high-bandwidth, low-fidelity assessments of incoming data coupled with low-bandwidth, high-fidelity processing (e.g., focus plus context techniques) are potential solutions for minimizing these problems.

• *Requisite memory trap.* People have a limited amount of short-term memory. As they gain experience with an environment (e.g., air traffic control), people can learn to be more efficient at building mental models, allowing them to form meaningful groupings of information and thereby extend what they can manage in short-term memory. However, the volume and complexity of information required for situation awareness creates situation awareness bottlenecks leading to memory failures with serious consequences. Techniques (e.g., information visualization, broadband displays) aimed at expanding human working memory capacity by offloading information patterns onto external memory may ameliorate these problems.

• *Workload, anxiety, fatigue, and other stressors.* People working in a disaster situation are particularly vulnerable to a number of serious stressors that tax their ability to maintain situation awareness. In a situation where lives are at stake, it is not surprising that stress and anxiety could significantly strain a person's ability to create and maintain situation awareness. Other factors may also come into play (e.g., self-esteem, career advancement). Stressors will undermine situation awareness by making the entire process of taking in information less systematic and more error-prone.

• *Data overload.* A person's ability to create and maintain a proper mental model of a situation can be overwhelmed by too much input. If more data are received than can be processed, situation awareness will quickly become outdated or contain gaps. While it is easy to think of this problem as simply a natural occurrence that people are ill-suited to handle, in reality it is often a function of the way that data are processed, stored, and presented in many systems. For instance, people are able to process graphical data much more rapidly than textual data (a picture is worth a thousand words).

• *Misplaced salience.* Salience is a measure of how compelling information appears to be. Yet what is most compelling is not necessarily what is most important. This is misplaced salience. The evolutionary process has adapted the human perceptual system to be more sensitive to certain signal characteristics than oth-

*continued*

> **BOX 2.2 Continued**
>
> ers. IT systems that take advantage of these adaptations can promote or hinder situation awareness. Unfortunately, many systems overuse knowledge of human salience, creating distractions that hinder the creation of proper mental models of situation awareness. People are unable to distinguish between false alarms and the *real* thing and begin to ignore or filter such input. Techniques for highlighting important information with pre-attentive codings, or presenting summaries appropriate to the task could help address this problem.
>
> - *Complexity creep*. Many IT systems are designed with complex sets of features that make it difficult for people to understand how they work or how to operate them. Videocassette recorders (VCRs) are a common example of this problem. Even automated flight management systems operated by highly trained pilots suffer from complexity creep. Complexity slows down a person's ability to correctly process and properly interpret information presented, ultimately undermining proper understanding of the situation. Training is often prescribed as a solution to dealing with complexity, but situations that occur infrequently (e.g., disasters) mean that a user is likely to have little or no experience with features used only in those situations and is highly likely to have forgotten or incorrectly remember the training.
> - *Errant mental models*. Mental models form a key interpretation mechanism for information taken in. They tell a person how to combine disparate pieces of information, how to interpret the significance of that information, and how to develop reasonable projections of what will happen in the future. Errant mental models cause people to misunderstand the situation. Mode errors (a special case of errant mental models) cause people to misunderstand information because they believe a system to be in one mode when it is in reality in another mode. This type of error is particularly insidious because it is difficult for people to realize their error. Research on IT to distribute more attention to highly diagnostic evidence and to search for disconfirming relations may provide a leverage point for addressing this problem.
> - *Out-of-the-loop syndrome*. Excessive or inappropriate automation can lead to this problem, where people are taken out of the decision-making loop and left with inadequate information about what the automation is doing or the state of the system. Automation can help situation awareness by reducing workload, but when automation reaches situational conditions it is not equipped to handle and the person has been left out of the loop, he or she may be unable to understand what is occurring and intervene in the situation.
>
> ---
>
> NOTE: The term "situation awareness" is used here, following the usage of the authors from whose work this material is adapted, rather than the perhaps more familiar "situational awareness" used elsewhere in the report.
>
> SOURCE: Adapted from Mica R. Endsley, Betty Bolté, and Debra G. Jones, *Designing for Situation Awareness: An Approach to User-Centered Design*, Taylor & Francis, London, United Kingdom, 2003, pp. 31-42.

specific questions."[14] There are a number of implications for the development of IT to aid human sensemaking, including integrated design of human interfaces, representational tools, and information retrieval systems. Sensemaking systems have both front ends (visualization) and back ends (content analysis and reasoning) that could aid human ability to skim, power read, recognize patterns, take notes, summarize, drill for details, and flag biases.[15]

Another factor that can color a person's understanding of a situation is organizational affiliation. Affiliation will drive decision making in early stages over situational needs. This reflects the need to fall back on ingrained and trusted culture and training when uncertainty is high and trusted information scarce. Over time, if uncertainty decreases and information on the situation increases, affiliation goal drivers dissipate and situational needs come to drive decisions. However, if there is the perception that the situation is out of control, then affiliation factors again drive the decision making. Situational awareness tools that help build a shared reality can reduce affiliation and culturally related misperceptions.

Given the number and variety of relevant parties, a detailed, accurate, and shared picture of both the disaster area and the status of the response is an obvious requirement for effective management of the disaster, and it is an area where information technology can be of considerable benefit. Technologies that can help in the near term include improved resource tracking (e.g., using radio-frequency identification tags [RFIDs]), information fusion from a priori known sources, reconciling data collection with privacy concerns, and publish-subscribe systems to deliver the information to the appropriate people. Longer-term research is needed to develop large-scale embedded sensor networks, automatic calibration of data confidence, automatic information fusion and data mining of diverse resources, routing of information to users based on semantics, filtering of false alarms, and effective presentation of information to users. Another relevant area is augmented cognition. DARPA is currently funding the Improved Warfighter Information Intake Under Stress program, which seeks to enhance human performance in diverse, stressful, operational environments by developing a closed-loop computational system in which the computer adapts to the state of the human to determine information presentation.

Data monitoring about situational variables, ranging from long-term

---

[14]See D.M. Russell, M.J. Stefik, P. Pirolli, and S.K. Card, "The Cost Structure of Sensemaking," *Proceedings of the SIGCHI Conference on Human Factors in Computing Systems*, Association for Computing Machinery Press, Amsterdam, The Netherlands, 1993, pp. 269-276.

[15]Mark Stefik, "The New Sensemakers: The Next Thing Beyond Search Is Sensemaking," *Innovation Pipeline* (a Red Herring newsletter), 2(10):13, December 2004.

variables (e.g., demographics) to midterm (e.g., status and location of stockpiles) to immediate (e.g., road and traffic conditions, weather), may be used to continuously update and adapt both mitigation and response plans to reflect local issues.

Finally, continuing advances in unmanned or robotic search and rescue capabilities could improve the ability to obtain access remotely to detailed information about casualties, infrastructure damage, and other information critical to response where responders are unable to reach as quickly or at all. Advances in this area hold promise for improving the safety of responders, as well as the timeliness and effectiveness of response and recovery.

## IMPROVED DECISION SUPPORT AND RESOURCE TRACKING AND ALLOCATION

Whereas situational awareness focuses on providing operators and decision makers with information relevant to their tasks and goals, decision support focuses on assisting them in formulating prospective actions, primarily by helping them understand and assess characteristics and consequences of alternative courses of action. Decision support is about explicitly recording candidate course(s) of action and generating, analyzing, and evaluating those alternatives. It also provides the means to monitor the effectiveness and progress of response activities. Lines between situational awareness and decision support tend to blur—decision support is dependent on understanding the situation, and decisions affect the subsequent situation, thus setting up a continuous feedback loop.

Some specific examples of how decision support systems might aid responders include the following: recommending on-the-fly decision evaluation, triggering "nagging" for decisions to be made and executed, tracking down the next alternate decision maker (an assistant chief, for example), tracking data and underlying simulation models used to make decisions against the actual situation and presenting warnings when deviations from the model appear, and updating the response plan. Further examples include the providing of early-warning triggers for notification that time is running out to exercise an option and the raising of red flags when decisions need to be made or when execution lags.

One open research challenge is how to support decisions in the face of significant uncertainty. Example research topics include computer-assisted decision-making tools, resource use modeling, risk management in the face of uncertain data and outcomes, sentinel processes to automatically monitor processes, and technologies that support distributed emergency operations centers. Progress on uncertainty management could yield significant progress—perhaps eventually making it possible to take actions that come well within 90 percent of perfect hindsight.[16]

An especially sensitive aspect of decision support tools deals with taking humans out of the decision loop. Some decisions may be appropriately made using a rule-based decision-making process. This could offload routine decisions so that human decision makers can focus more attention on those decisions requiring policy-based decision and judgment. Determining which decisions can be rules-based and which policy-based and under what circumstances will require both research and an adoption strategy that allows disaster managers time to build trust and confidence in these systems.

Decision support technology could also facilitate access to and collaboration with other organizations, such as mass media, the private business sector, the Army Corps of Engineers, public health and public works, as well as traditional first responder organizations.[17] Indeed, the global character of current information and communications technology means that the decision support could come from qualified sources anywhere in the world.

The issue of scalability requires particular attention for decision support systems to be truly effective in a large-scale disaster. For disaster management, scalability of these systems is intertwined with the issue of the dynamics of the situation. IT has been implemented on a massive scale, resulting in impressive gains in efficiency and productivity. But the inherent chaos that arises in a disaster creates unique problems for realizing these gains in disaster management practice.

Closely linked to decision support is the topic of logistics management. The ability to monitor movements of personnel, goods, services, and victims; to recognize mismatches; and to trigger adaptive action (whether it be redirecting the patients or augmenting the facility) would greatly enhance disaster response management. Often, retrospective analyses of disaster responses effectively say, "Regardless of resource limitations, we didn't make the best use of the resources that we did have."

Routing systems for trucking and airline industries are examples of

---

[16]Laboratory results are already beginning to show early promise toward this possibility: e.g., see R.T. Maheswaran, C.M. Rogers, R. Sanchez, and P. Szekely, "Reward Estimation by Communicating Aggregated Profiles (RECAP): Distributed Coordination in Uncertain Multi-Agent Systems," submitted to 20th International Joint Conference on Artificial Intelligence, Hyderabad, India, January 6-12, 2007.

[17]For more perspective on collaboration and associated tools, see C.A. Bolstad and M.R. Endsley, "Choosing Team Collaboration Tools: Lessons Learned from Disaster Recovery Efforts," *Ergonomics in Design* 13(4):7-13, 2005.

decision support systems that manage the reallocation of resources and give human operators feedback on potential options. These systems know the resources available and make adjustments based on changes in the status of those resources. They handle massive amounts of resources and processes in a highly efficient manner—they scale extremely well—as long as resources are added slowly and in an orderly manner and processes are predictable. They are optimized for bounded, stable situations. They handle problems by planning for a degree of reserve capacity. These systems become quickly overwhelmed when reserve capacity is exceeded. The problem of decision support for disaster management is that it is inherently an out-of-bounds, unstable situation. (See Box 2.3 for a discus-

---

**BOX 2.3**
**The Limits of Commercial Logistics Operational Models for Disaster Management**

A frequently expressed frustration with recent disaster management efforts concerns the inefficiency of logistics operations deploying resources (e.g., ice, water, trailers, medical supplies, generators) to affected areas. "If FedEx and UPS can do it, why can't disaster managers?" is a common refrain. As performed today in disaster management practice, logistics activities tend to make very limited use of information technology (despite the models provided by leading-edge distribution companies).

Recent investments by the Federal Emergency Management Agency have begun to bring its practices up to date with those of commercial systems, adding some tracking capabilities, though it appears that much more can be done.[1] Expanding these investments will certainly improve logistics operations. Ideally, technologies like radio-frequency identification (RFID) tagging, Global Positioning System (GPS), and others used to manage massive supply chains like those managed by Wal-Mart and others could be pervasively implemented for disaster management.

However, there are critical differences between commercial logistics operations and those required for disasters. These differences mean that even if state-of-the-art commercial logistics technologies are adopted, they will go only so far in addressing the logistics and resource management needs in a disaster. The importance of techniques for recognizing impending problems is not generally understood. Thus, the history of military and emergency supply chain management contains periodic calls for blindly adopting commercial methods. Rapid shipping companies such as FedEx are quite rightly lauded for the speed and efficiency of their operations. However, the underlying algorithms are designed not to ensure that all packages are delivered on time, or even that important packages are delivered on time. Priorities are handled through channeling into faster shipping methods, but the same algorithms apply. The underlying commercial algorithms determine the schedules and transportation resources needed to ensure that *most* packages will be delivered in timely fashion *if* the depots and distribution centers do not change *and* the situation stays within predicted bounds. The underlying algorithms further assume that priorities of packages are undifferentiated within broad categories (i.e., if you cannot deliver all packages in a priority class, it is

## BOX 2.3 Continued

acceptable to allow circumstances to determine which ones get through). They also tend to assume that packages' priorities do not change once in the system and that destinations of packages do not change once in the system. If *any* of these assumptions is violated, commercial shipping organizations do the cost-effective thing: they apologize, remind customers of disclaimers in the shipping agreement, and perhaps offer a refund.

The techniques used by commercial organizations are highly adapted to their purpose. They have achieved excellence by developing specialized logistics solutions, but to problems that are qualitatively different from those present during disasters. Ensuring delivery of critical items in disaster situations is a related problem—but it is not the same problem. Because it is a related problem, techniques that have been applied to the commercial problem are relevant—but recognizing that they are relevant is not the same as having actually adapted and applied them. Nor, because the problem is related but different, are those techniques sufficient. When lives depend on a shipment—deaths ensue if that shipment is late—the function, and thus the underlying algorithms, must be different.

Disasters are by definition unstable, "out-of-bounds" situations. Logistics systems developed for them must be designed to handle a complex and evolving set of problems. Once problems with a plan or schedule are detected, the critical question is how to revise it to contain the ripple effects—which is essential to minimizing costs and time delays in adapting to the changed situation.

Thus, a number of technical challenges differentiate the commercial problem from the disaster management problem:

1. Scale introduced by the surge requirements inherent in disasters,
2. Complexity introduced by the much-finer-grained and more strongly interacting issues of prioritization,
3. Complexity introduced by dynamics (managing the risk that actions will be rendered inappropriate by the fluidity of the situation),
4. Complexity introduced by information uncertainty (managing the risk that actions will be rendered inappropriate because they were based on incomplete or inaccurate information), and
5. Minimally disruptive plan revision.

Because of the inherent complexity of the number of issues and constraints that must be considered, humans are limited in their ability to handle these challenges unaided by technology. Instead, for ordinary situations, most approaches have focused on maintaining reserve capacity as a buffer. A large amount of work in operations research and other fields has focused on techniques for calculating what a safe reserve capacity should be. Disasters, by their very definition, exceed available resources and overtax the process of effectively allocating them. Thus, computing and maintaining safe reserve capacities, although highly necessary, is not sufficient. Intelligent adaptive planning technology that addresses the five challenges listed above is still largely in the research stages. Nevertheless, it is a critical capability.

---

[1]Robert Block, "FEMA Regroups After Katrina, But Some Question Its Readiness," *Wall Street Journal*, August 7, 2006, p. A1.

sion of differences between commercial logistics operations and logistics and resource tracking for disaster management.)

As performed today in disaster management practice, logistics, resource tracking, and allocation activities tend to make very limited use of information technology (despite the models provided by leading distribution companies). Consequently, any notification of problems is likely to be reactive rather than proactive (e.g., "We're at the bridge but it's down. . . . Weren't they supposed to be here by now? . . . Why are there only 500 child-sized crutches when we asked for 5,000?").

Problem notification and consequent situational awareness tend to be sequential and therefore subject to propagation delay during which parties can get out of synchronization, rather than having immediate and shared situational awareness. (An all-too-familiar conversation frequently starts with Destination asking Dispatch, "When are they coming?" and Dispatch replying, "They're not there? I'll find out what's going on and get back to you."). Problem resolution processes are unlikely to identify and simultaneously involve all affected parties, thus producing local instead of global solutions.

Existing technology makes it quite possible to do much better. With RFID tags, every item, box, pallet, and truck can know and report its relationship to the shipping manifest. With the Global Positioning System (GPS) and satellite connections, continuous position reporting is possible. Thus, knowing what you have, exactly where it is, and where it is going (even if the destination itself is a moving target) is all possible. Geographic information systems make it possible to record and display that status information, while simultaneously recording and displaying other status information such as weather conditions and transportation infrastructure condition.

It is easy to set alarms when planned directions of travel and rates of motion are not met, and it is feasible for users monitoring these systems to notice when displays indicate potential barriers to planned movements. When human operators notice problems, route-planning software can greatly facilitate rerouting the movements. Peer-to-peer file-sharing tools can ensure that all participants automatically receive updates of the information they need. Many such commercially available tools have encryption built in, ensuring the security of shipment information in the event of extreme situations in which theft or diversion is a concern. Exercises such as Strong Angel have demonstrated the effectiveness of relatively crude measures to provide synchronization, even in situations where conventional communications have broken down (e.g., jeeps or unmanned aerial vehicles [UAVs] surveying an area and receiving and re-broadcasting update information on wireless network frequencies).[18]

---

[18]See Strong Angel III at http://www.strongangel3.net.

Thus, solutions exist today that would reduce error rates, promote proactive action in recognizing problems, and provide higher levels of shared situational awareness regarding those problems. Functions that remain problematic, however, include initial decisions about resource allocations, recognition of many problems in the execution of plans and schedules, and development of good repairs to the plans.

If these problematic issues are addressed, it is possible to do even better. There are nascent technologies in the research community that could make a significant impact. If nurtured and transitioned, these would produce far better initial resource allocations, identify a wider range of problems proactively (with significantly greater lead time for resolution), and help develop more effective and globally beneficial solutions to problems as they arise. Effective determination of needs requires combining pushes and pulls—interrelating, rather than stovepiping, intelligent forecasting of requirements with rapid aggregation of requests.

Simulation systems provide one useful tool for decision makers to test potential resource allocation and planning options in a virtual environment. They can provide a vehicle to promote understanding and dialogue on actions and issues related to the development of an effective preparedness and response plan, and serve as a forum and basis for mutual understanding between agencies and disaster management practitioners. Further advances in simulation environments promise to provide comprehensive modeling frameworks that integrate both inverse and forward points of view, applicable at multiple levels of analysis in diverse fields of study, in a structured manner. Computational architecture that is flexible, scalable, and adaptable promises the ability to create persistent virtual worlds for continuous replication, verification, validation, uncertainty quantification, and margin-of-error estimations.

Alternative recovery plans could be developed and tested in the context of simulations and risk models that allow plan effectiveness to be tested and that enable continuous adaptation of plans in light of available resources and past experience. Instrumentation and data collection are critical elements for learning from one disaster to amend management practices of future events. Data collected during response and recovery operations can also be used in post-disaster analyses to feed future mitigation efforts. It can be used to validate and improve models and simulations. Learning from instrumentation and post-incident analysis have proven invaluable in other fields, such as health (with today's emphasis on evidence-based medicine), defense (where the military collects reports and other data from exercises and operations to develop lessons learned that are fed back into the development of both doctrine and weapons systems), and air transportation (where the National Transportation Safety Board investigates every civil aviation accident in the United States in order to improve the reliability and operational safety of airplane systems). Indeed, in most of these fields incorporation of lessons learned is highly systematized to ensure timely implementation of changes in practices and systems.

Advances in high-performance computing are now demonstrating the ability to execute, log, and analyze discrete event simulations with literally millions of separate ongoing computational processes. These have been integrated into wargames extending over weeks, involving tens and hundreds of thousands of human personnel. This means that it is becoming technically feasible to run situation analysis systems for disasters that continuously operate on "best available understanding"—filling in missing information with simulations, models, and forecasts when necessary, replacing them with sensor data, situation reports, and incoming supply requests when available.

Advances in computing power, as well as the development of new algorithms, also create the prospect of adaptive planning, scheduling, and resource allocation processes that continually fine-tune logistical support plans to the evolving situation. These algorithms are increasingly able to reason about managing the risk of a plan in the face of various kinds of uncertainty. One type of uncertainty concerns information quality. (How sure are we that the roads are passable?) Another type concerns the likelihood of success for potential courses of action. (Since we don't know whether a convoy from the north will get through quickly, would it make sense to also route some of the supplies from the south or to send excess shipments that will be redirected if the others get through?)

In addition, advances both in power and accessibility of high-performance computing combined with new algorithms are pointing the way toward automated solutions of increasingly larger planning and scheduling problems. Not only are the algorithms more efficient, but techniques such as "backdoors" and hybrid problem solvers have shown how large problems can be transformed into simpler and more manageable ones. "Backdoors" is a technique for identifying critical decisions within a large set of alternatives that, once made, reduce the number of other options that have to be considered without significantly reducing the likelihood that the best overall choices are still made. Hybrid problem solving is a related technique in which fast but "suboptimal" techniques are used to produce rough sketches of plans, which are examined for characteristics that can be used to restrict the options considered by slower, higher-quality algorithms.

Taken together, these techniques offer a future in which much more efficient plans are generated on the basis of much more meaningful information. This ensures that responses are initiated with the best-tuned courses of action available at the time. The next challenge is to keep those responses tuned when the situation changes during the course of executing them.

In order to do better, the challenge is to provide individuation of shipped items. The first element of this, tracking the items, is increasingly easy (see above). The next challenge is to use the tracking information to trigger warnings that a shipping plan is going awry. As noted above, the technology is basically in place to do this by asking, "Is it where it should be at this time?" However, it is possible to do better than that. If a bridge on the trucking route is down, rerouting should start as soon as any part of the system knows that fact—rather than waiting for an overtaxed human to notice, much less waiting until the trucks get to the bridge.

DARPA has funded related research in the areas of plan sentinels and mathematical approaches for estimating progress and probability of success on goal structures. Plan sentinels extend a range of existing planning techniques that use explicit descriptions of goals and methods for achieving those goals. Plan sentinels focus on capturing assumptions underlying methods and generate software in order to explicitly monitor data streams for evidence that those assumptions are violated. For example, driving a truck from point A to B is a method for getting to a location which assumes that roads and bridges are intact, so plan sentinel software would generate code for monitoring data streams and routing traffic reports. In contrast to plan sentinels, the mathematical approaches adapt techniques such as nonlinear filtering methods (previously used to track progress in geometric spaces) in order to track progress in symbolic spaces. These techniques generate probability distributions that can be used to assess likelihood of the "speed and direction" of future progress on the plans.

Symbolic approaches like plan sentinels and mathematical techniques such as just described can make complementary contributions to the ability to proactively predict problems and to begin as early as possible to respond to them.

## GREATER ORGANIZATIONAL AGILITY FOR DISASTER MANAGEMENT

Disasters are varied and no single organizational structure is necessarily the best for dealing with the range of possible incidents. During disaster response, organizations must be able to form quickly and work well together across space and time. They must be able to adapt and resize easily as the disaster develops. The 9/11 disaster expanded in scope and scale in a matter of hours even as the primary emergency operations center was lost within the World Trade Center complex. The National Response Plan and National Incident Management System provide frameworks for dynamic organizations to be formed, but do not address the diversity of technology in different organizations, the lack of rapport, or the ability of organizations to quickly integrate operations. They also articulate an elaborate set of command and control systems that are inflexible and work against the idea that agility and flexibility are at least as important as command and control hierarchy.

Simulation systems offer one avenue for training disaster management professionals, from first responders to disaster managers, to better anticipate problems and to become more flexible and adaptable to working across organizational boundaries, integrating operations, and adapting to different technologies. Simulation systems can simultaneously interface with and drive both planning systems and training systems.[19] This would enable preparatory work to ensure the robustness of plans against multiple scenarios. It could also support training according to the plans and seamless transition into systems for response execution. Using detailed analysis of the response to past disasters, these simulations would be solidly grounded in reality, and their accuracy would increase with each incident.

Capturing lessons learned and making them available in a form useful to the broad community of disaster management is another area of potential for IT to support organizational learning. Learning from past experiences is an important element supporting the ongoing improvement of disaster management. Lessons learned can cover a wide range of topics—from experience with specific processes and organizational structures to the effectiveness of specific communication technologies or software systems. The Department of Homeland Security has established Lessons Learned Information Sharing (http://www.llis.gov) as a mechanism for disseminating lessons learned.[20] It provides an illustration of how the Internet can be used to support dissemination of lessons learned. Nevertheless, lessons learned from previous disasters have not typically propagated quickly through the disaster management community.

Another organizational issue is management of the flow of personnel in and out of an incident area. Improved IT infrastructure for credentialing and identification checks—who they are, whether they have the capabilities they say they have—could improve the efficiency of response and recovery operations, such as dispatching medical workers, repair technicians, and other appropriate people into a disaster area. In the aftermath of Hurricane Katrina, much damage to information and communications

---

[19]Several areas of potential opportunity to improve disaster management practice using simulation technology are drawn from a briefing to the committee by Alok Chaturvedi, director of Purdue University Homeland Security Institute. Presentation on September 20, 2005.

[20]Lessons Learned Information Sharing is a national network for emergency response providers and homeland security officials to share lessons learned and best practices. For more information, see https://www.llis.gov.

infrastructure was fixable, or there were backups and alternatives. However, the lack of access because of the security ring around New Orleans and limited means for authorities to quickly validate people's credentials meant that repairs could not be effected; generators ran out of fuel and alternatives could not be switched in. Similar problems involving physical access were cited during the response to 9/11.[21] Recent credentialing efforts for emergency responders, including databases to keep track of volunteers, have been advancing and expanding to include telecommunications specialists, utilities workers, and other private-sector disaster response workers.

Authentication and credentialing constitute a complex topic, involving many technical and non-technical issues.[22] A few example areas of where further IT research might bear fruit include voice print analysis in the network, fingerprint sensor on the push-to-talk radio buttons,[23] RFID tags (badges), and verbal authorization codes issued to first responders and other authorized personnel (e.g., city workers, volunteers, and so on) that are linked to the radios. In addition, back-end database and architectural considerations regarding the identity and credentialing management system as a whole also merit attention. Currently, as with Internet purchase orders, identity authentication is accomplished through the verification of several independent pieces of information, not one code or password. In the case of false negatives, lost tags, mismatches between tag and radio, and so on, one model of mitigation is for the system to transfer the transaction to a dispatcher who could decide whether sufficient evidence exists to authorize authentication. The entire realm of credentialing and identity management is clearly a fruitful research area.

Further advances are needed to improve understanding of how communication structures map onto organizational structure requirements. Dynamic authority mapping is one potentially useful tool.

In the short term, information technology is likely to be applied chiefly to automate and accelerate traditional disaster management processes and practices. In the midterm to long term, however, increases in information-processing capacity offer the potential to enable a transfor-

---

[21]See, for example, National Research Council, *The Internet Under Crisis Conditions: Learning from September 11*, The National Academies Press, Washington, D.C., 2003.

[22]More information about the technical, architectural, and policy challenges associated with large-scale identity and credentialing systems can be found in a report from the Computer Science and Telecommunications Board; see National Research Council, *IDs—Not That Easy: Questions About Nationwide Identity Systems*, National Academy Press, Washington, D.C., 2002.

[23]One potential drawback is that such a radio could not be used by others if the designated operator was disabled or otherwise not available.

mation of those very processes by enabling innovation in organizational practices. Such IT-driven shifts have occurred in many sectors, with major organizational implications.[24] Although these transformations are not entirely predictable, the empirical evidence suggests a number of possibilities.

The military's hierarchical command chain is one well-known organizational model for managing extremely complex, distributed activities. It reflects the development over many years of a clear sense of what to centralize and what to decentralize. In recent years, the military has undertaken significant, IT-enabled revisions to doctrine based on the idea of providing more information to individual units or warfighters and enabling increasingly distributed network-centric operations. The committee believes that a similar analysis and evolution of doctrine that takes into account the unique characteristics of disaster management (such as diverse actors and jurisdictions in a federal system and the important role of private-sector organizations) as well as new technological capabilities are also needed.

One such possible shift would be from information-centered architectures updated in batches (e.g., "reports") toward distributed processing of continual messaging-streams fed by pervasive sensors providing real-time situational awareness data, with different users detecting trends and transitions according to their local requirements. There could be a corresponding shift from specialized management systems that are activated for disasters and deactivated afterward to an "always-on" state of activation that varies more in scale than in nature as events occur. A move should also be possible from command models of resource management toward negotiated "brokerage" approaches working with current models of the best actions that can be taken with available resources. Another possibility is the ability to reach a mature compromise between the dual extremes of rigid bureaucracy and all-to-all interoperability toward a "managed ad-hoc-racy" of disaster management and responder organizations that can evolve seamlessly and continuously over the entire course of a disaster. Finally, a role-based concept of individual and unit identity could reduce the significance of jurisdictional, disciplinary, and even official/civilian distinctions. The cumulative effect of these changes could be a shift from a mechanical focus on preserving and restoring the status quo ante toward a flexible strategy of resilience and adaptability to the dynamics and inherent complexities of disasters.

A common organizational theme is the strong tension between cen-

---

[24]Jason Dedrick, Vijay Gurbaxani, and Kenneth L. Kraemer, "Information Technology and Economic Performance: A Critical Review of the Empirical Evidence," *ACM Computing Surveys* 35(1):1-28, 2003.

tral authorities, which want to assert hierarchical control over disasters, and the distributed nature of most disasters. Authorities may want to be seen as "in charge," even though most of the actual work in disasters results from the many less-coordinated and distributed actions of individuals. Responders typically bring tremendous energy to the scene. One response is to put someone in charge to channel (and bind) this energy. Another is to let the energy emerge and then harvest it. Organizations have a hierarchical comfort zone that has driven them to the former response, but disasters are also accompanied by the rapid development of emergent multiorganizational networks.[25]

These networks form the locus for collective sensemaking and organizational learning under conditions where ambiguity and uncertainty are an inherent part of the environment.[26] Information technology could support emergent networks by helping them deal more effectively with unpredictable information sources, lowering barriers to information flow, making organizational boundaries more permeable, and easing coordination between diverse and distributed actors. For example, a number of emergent groups, existing in entirely virtual space and formed using the Internet and technology such as blogs and wikis, performed important services during Hurricane Katrina. Research to find more systematic ways to leverage such technology may yield new means for supporting emergent networks.

While this report is primarily limited to the IT aspect of disaster management and did not specifically look at problems from an organizational theory or management theory point of view, the potential contributions that management and organizational science can make to better understanding the situation in command centers and to improving other aspects of disaster management are undoubtedly significant. Moreover, as research progresses in these areas, analysis both of the impacts that greater use of IT may have and of how IT can help ameliorate other problems will be useful. For instance, organizational research on collective sensemaking could help focus IT research on how to address confusion that stems from ambiguous information rather than simply finding better ways to reduce ignorance arising out of uncertainty by increasing the quantity and quality of information.[27]

Another common theme, and one that came up frequently in testi-

---

[25]Kathleen Tierney and Joseph Trainor, "Networks and Resilience in the World Trade Center Disaster," *Research Progress and Accomplishments 2003–2004*, Multidisciplinary Center for Earthquake Engineering Research, University at Buffalo, State University of New York, May 2004, pp. 157-172.

[26]K.E. Weick, *Sensemaking in Organizations*, Sage Publications, Newbury Park, Calif., 1995.

[27]Ibid., pp. 185-187.

mony to the committee, is the issue of building trust among the various actors involved in responding to a disaster. But in a disaster situation, cooperation is often required between people who are strangers with no existing trust relationship. Thus, approaches and supporting technologies are needed that aid coordination and structure across people and organizations that have little trust but some common goals. IT's ability to enhance organizational agility in disaster management may be limited by the extent to which it is unable to overcome barriers to working in an environment of limited trust. Applying existing understanding about the relationship between trust and technology and extending that knowledge through further research will be critical to advancing organizational agility.

## BETTER ENGAGEMENT OF THE PUBLIC

Two distinct aspects of better engagement of the public through better use of IT involve (1) the use of warning systems and broadcast alerts to inform the public of actions that they should take to protect themselves and their property and (2) the ability to leverage the public as providers of information and sources of valuable technology tools. The potential to improve the use of IT in both areas is substantial, although the second will also require considerable shifts in culture among public safety and emergency management professionals.

### Alerting and Warning Systems

Improving warning systems for various types of disasters has received considerable attention, especially in the aftermath of the 2004 Indian Ocean tsunami.[28] Effective warnings save lives, reduce damage, and speed recovery. Warnings are most effective under the following circumstances:

- They are accurate and result in appropriate action.
- Any probabilistic aspects (e.g., likely hurricane landfall probabilities) are clearly communicated.
- They are standard, consistent, and easily understood.
- They are delivered to just the people at risk and in a timely manner.
- They are delivered through a variety of mechanisms to achieve maximal reach.

---

[28]Subcommittee on Disaster Reduction, *Grand Challenges for Disaster Reduction*, Executive Office of the President of the United States, Washington, D.C., June 2005.

Technology has greatly improved the ability of forecasters to make accurate predictions about natural disasters. Public education has improved actions that people take in response to warnings. Experience and policy changes have made authorities better at communicating consistent, clear messages. Further improvements are possible in all of these areas—especially through broader deployment of sensor systems and further advances in sensor technology. But, there is now a significant gap in how warnings are delivered and what is possible with existing technology.

Forecasting and sensing technology has made it possible for siren-based warnings to be issued minutes to seconds before the onset of a disaster where previously no warning was possible, and future advances may improve detection and lengthen warning times (earthquake detection is one such promising area). Where the time for delivering alerts is still very short, sirens can be highly effective because they can be rapidly and broadly disseminated. Sirens require people to know what action to take (e.g., for a tornado find shelter, under ground if possible). Public education and drills are used to instill this knowledge. State-of-the-art siren technology can now include a public address capability, allowing more specific information to be communicated. However, sirens are inherently outdoor systems. They are valuable because they may reach people who would otherwise receive no warning. They will continue to play an important role in alerting the public, and technology advances can make them even more effective.

Warning systems using information and communications networks can be significantly upgraded using existing and emerging technologies. The Congressional Research Service report *Emergency Communications: The Emergency Alert System (EAS) and All-Hazard Warnings* describes a number of government efforts to develop a digital warning system, including the ongoing pilot projects of the Federal Emergency Management Agency, the Information Analysis and Infrastructure Protection directorate at DHS, and the Association of Public Television Stations to develop an integrated public alert and warning system.[29]

A Presidential Executive Order of June 26, 2006, aims to establish an integrated alert and warning system and to "establish or adopt, as appropriate, common alerting and warning protocols."[30] Revamping the Emer-

---

[29]Linda Moore and Shawn Reese, *Emergency Communications: The Emergency Alert System (EAS) and All-Hazard Warnings*, Congressional Research Service (CRS) Report for Congress (RL32527), CRS, Washington, D.C., 2005.

[30]George W. Bush, *Executive Order: Public Alert and Warning System*, 2006; available at http://www.whitehouse.gov/news/releases/2006/06/20060626.html.

gency Alert System (EAS) was also a finding and recommendation in the Federal Communications Commission (FCC) independent study of communications during Hurricane Katrina[31] and in the White House report on Katrina.[32] These reports note the potential of new technologies (satellite, cellular, pagers, Internet, wireless) to send more targeted messages. As the FCC review of EAS notes, "wireless products are becoming an equal to television and radio as an avenue to reach the American public quickly and efficiently."[33]

One common technological denominator in recent efforts has been the Common Alerting Protocol (CAP), a warning format standard developed by emergency managers, promoted by the Partnership for Public Warning, and codified by the Organization for the Advancement of Structured Information Standards (OASIS) standards organization. CAP has been used in most of the major warning system prototypes in recent years and features prominently in the FCC proceedings on the future of EAS.

## Communication with the Entire Affected Population

The importance and challenge of reaching the entire affected population—including all social and socioeconomic groups, disabled, elderly, and other special-needs groups—before, during, and after a disaster strikes were highlighted by Hurricane Katrina. More generally, reaching these groups is among the most significant of issues relating to engaging the public to ensure their own survival and recovery during disasters. Hurricane Katrina tragically demonstrated the error in assuming that better communications alone guarantees effective public engagement.[34] It also served as a reminder that access to and familiarity with information technology is not universal. Further, the ability to act on available information, even when it is accessible, may be limited.

---

[31]See Federal Communications Commission (FCC), "Review of the Emergency Alert System, First Report and Order and Further Notice of Proposed Rulemaking," FCC 05-191, Washington, D.C., November 2005; and FCC, *Recommendations of the Independent Panel Reviewing the Impact of Hurricane Katrina on Communications Networks*, FCC 06-83, Washington, D.C., June 2006.

[32]See The White House, *The Federal Response to Hurricane Katrina: Lessons Learned*, February 2006, pp. 83, 109-110.

[33]FCC, "Review of the Emergency Alert System: First Report and Order and Further Notice of Proposed Rulemaking" (EB Docket No. 04-296), FCC, Washington, D.C., p. 29; available at http://hraunfoss.fcc.gov/edocs_public/attachmatch/FCC-05-191A1.pdf.

[34]National Research Council, *Facing Hazards and Disasters: Understanding Human Dimensions*, The National Academies Press, Washington, D.C., 2006, p. 68.

## Risk Communication

Understanding what information should be provided to the public before, during, and after a disaster falls under the general topic of risk communications, which an earlier National Research Council report defines as "an interactive process of exchange of information and opinions among individuals, groups, and institutions [and] a dialog involving multiple messages that express concerns, opinions, or reactions to risk messages or to legal and institutional arrangements for risk management."[35] Insights from such work can be used to inform future efforts to apply IT for improved risk communication.

## The Public as an Information and Technology Resource

While warning and alerting systems are important, their perspective is of the center talking out to the masses. The committee finds vast potential to further engage the public by changing this perspective to embrace two-way communications between authorities and the public. Interactions with the public are an important part of disaster management, yet these interactions have received relatively little attention. Changes in technology available to the public mean that there are not only new ways to reach them (with warnings), but also new ways for the public to both gather and communicate information. One typically thinks of the public playing a passive role of information receiver in a disaster. People have thus far been engaged only marginally and conditionally as sources of information with valuable technology in critical or otherwise unreachable locations (e.g., certified amateur radio operators). The now ubiquitous 911 emergency calling system is an example of how responders can get useful information from the public. It is a simple mechanism for allowing the public to report information.

Yet, there is potential for the public to play a much larger role during a disaster, and information technology is increasingly making it possible to engage the public in a variety of ways. Civilians—people on the street—are nearly always the very first people on the scene of disaster, especially in situations with little or no warning. Collectively they have a richer view of at least a small portion of a disaster situation than is available from within an emergency operations center. Even 10 years ago most people carried little or no technology around with them. Today, much of

---

[35]National Research Council, *Improving Risk Communication*, National Academy Press, Washington, D.C., 1989.

the public has sophisticated mobile communication and sensing capability. Camera phones, wikis, the Web, and text messaging are all capabilities increasingly available to people on the scene. Harnessing these sources of valuable information holds great promise for providing critical information to disaster managers, especially in the initial response stages.

Victims can be transformed into actively engaged responders if given a meaningful and appropriate means to participate. They can be recruited to assist in disaster response. They can be kept informed as to how and when to act appropriately. Furthermore, they can offer critical redundant IT resources as traditional sources are impacted by a disaster. IT mechanisms that interface disaster response agency information systems to interactive public communications channels (e.g., Internet, wireless communication) could provide information gathering and dissemination mechanisms that ameliorate problems of agency overload from affected populations seeking situational information or having the ability to provide local information.

The redundancy of this approach would also improve the reliability of the communications with the attendant advantages of improved performance and public perception of appropriate actions. Valid concerns about the trustworthiness of information have inhibited any major steps toward incorporating these types of changes more fully. Yet, changes in the amount and quality of technology carried by individuals and continuing advances in filtering, qualifying, and analyzing information of uncertain quality means that a major opportunity may be missed for making better use of IT, especially given limited resources available to public officials.

The Strong Angel exercises have explored "techniques and technologies that support the principle of resilience within a community" with the objective of effectively tapping "the expertise and creativity within an affected community, including through public-private partnerships. A second overarching objective is the development of social tools and techniques that encourage collaborative cooperation between responders and the population they serve during post-disaster reconstruction."[36]

Cybercitizen networks also hold promise as an important new element of response similar to, but on a much larger scale than, the role filled by amateur radio operators. IT can help facilitate these interactions. In the near term it is possible to use or adapt existing technologies to create validated online information sources, to deploy multimodal public reporting sites, and to build directories of private resources to facilitate

---

[36]See Strong Angel III at http://www.strongangel3.net.

deployment. Research is needed into technologies that do dynamic capability profiling and credentialing, semantic routing, and filtering of multimodal public input, and research is also needed to optimize data formatting for diverse terminal devices.

## ENHANCED INFRASTRUCTURE SURVIVABILITY AND CONTINUITY OF SOCIETAL FUNCTIONS

Disasters inherently cause disparate communities, infrastructures, and organizations to interact (and not interact) in unanticipated ways. Hurricane Katrina is an obvious example, having displaced much of an entire metropolitan region, with residents being dispersed across the country. But smaller-scale disasters also disrupt societal functions. Families and friends are separated. Families need access to housing, schools for children, and social services. People lose access to medical facilities and medical records. Jobs are lost. Important cultural institutions are disrupted. There is considerable opportunity for improving the use of IT to reconnect people, provide a temporary bridge to restore and maintain relations and interactions, and to speed their restoration. Hurricane Katrina demonstrated that people can come together, using the Internet and other information and communications technologies, to apply knowledge, skills, and technology to have a positive impact on the lives of those affected by a disaster.[37]

Another well-publicized example of an emergent use of IT was the creation of wikis[38] that enabled volunteers to connect with victims. Some of the functions wikis served in the aftermath of Katrina were listing helpline numbers, posting offers of temporary shelter, identifying where and how to make donations, serving as a clearinghouse for identifying government resources, offering health and safety information, sharing advice and experience on relocation, publicizing fund-raising events, providing information about lost and found pets, reconnecting families, and posting help-needed notifications. The emergent behavior exhibited by these wikis is one of their great strengths—rising to meet an unanticipated and highly diffuse need. They are highly flexible and adaptable and

---

[37]For example, see Keith Axline, "Craigslist Versus Katrina," *Wired News*; available at http://www.wired.com/news/planet/0,2782,68720,00.html.

[38]According to Wikipedia, "a wiki is a type of website that allows users to easily add, remove, or otherwise edit and change some available content, sometimes without the need for registration. This ease of interaction and operation makes a wiki an effective tool for collaborative authoring. The term wiki can also refer to the collaborative software itself (wiki engine) that facilitates the operation of such a website." See http://en.wikipedia.org/wiki/wiki.

demonstrate just one way that the emergence and ongoing advance of collaborative tools could contribute to disaster management. The challenge for disaster managers is to leverage the power of these emergent uses of IT—and support research and development to advance their application to disaster management—without destroying their inherent flexibility and adaptability.

In addition to making better use of IT to help people, IT can be used to improve the survivability of critical infrastructure—another major factor in speeding recovery and restoring societal functions. The committee heard numerous possibilities for advancing IT and extending its applications to improve the resilience and management of critical infrastructure systems, such as the electric grid, water, transportation, housing, and health care. The interdependencies between these systems are often not well understood and rarely proactively engineered for resilience because they are usually designed and operated by independent entities over a long period of time. The structural couplings between these systems may also manifest themselves on a wide range of spatial and temporal scales, making it difficult to quantify them. Solving these problems requires different jurisdictions—cities, counties, states—to work closely with each other and with federal agencies. The restoration of New Orleans, for example, is widely understood to require a concerted rebuilding across government agencies, public safety organizations, businesses, and public utilities. Simply clearing damage, processing insurance claims and other compensation, and rebuilding residential, government, and commercial infrastructure constitute a very complex, multifaceted process that is likely to take years to complete.

Advances in IT can revolutionize other technical disciplines with direct and indirect implications for advances in disaster management. A salient example is the entirely new class of monitoring and control capabilities made available to civil and mechanical engineers by the creation of small sensors, microprocessors and wireless communication devices. Many applications require the deployment and use of sensors on a wide scale—capabilities that are starting to emerge from research into distributed sensor networks.

When terrorists attacked the World Trade Center in New York in September 2001, thousands of occupants of those doomed structures had an hour to escape. But delays in the assessment of the structures' integrity stole crucial minutes from the evacuation and ultimately doomed thousands of innocent civilians. Today it is possible to instrument such structures using sensors and wireless connections, in such a way that the changing forces within the towers' structures could be recognized and evaluated almost instantly.

On a larger scale, ubiquitous monitoring of the condition and utilization of highways could trim minutes or even hours from the travel time of responders by routing them around damaged or clogged routes. Supervisory Control and Data Acquisition systems for water, gas, and other utilities are being enhanced to provide detailed analysis of damage due to earthquakes or explosions, enabling system operators to speed restoration and minimize peripheral disruptions of service.

The benefits of comprehensive monitoring and management of engineered systems can extend beyond their own boundaries, for example managing interactions between systems, such as the power grid and the communication networks that rely on them. This underscores the importance not only of collecting system-specific data but also of normalizing and exchanging real-time assessment data between systems.

Buildings, roads, and other constructed infrastructure exhibit significant resilience and robustness in the face of disaster. However, infrastructure that appears to be intact may in fact have been severely damaged in ways that are not readily apparent. For example, in the wake of an earthquake, a building might be on the verge of collapse or a bridge might be ready to fail after even the smallest aftershock. By making hidden damage more apparent, sensors combined with information technology can enhance response and recovery operations by reducing uncertainty about the state of infrastructure.

Continuous monitoring and analysis of critical infrastructure could be done by developing new instrumentation capabilities. This would enable the routing of sensor information from buildings, bridges, and infrastructure systems—for example, roads and water, gas, sewer, communications, and power systems—to monitoring locations, providing responders with information about the robustness and safety of the infrastructure affected. As in other areas, power supply independent of the electric grid is a critical issue that must be addressed to extend sensor capabilities.

# 3

# Improving Acquisition and Adoption of IT for Disaster Management

This chapter focuses on information technology (IT) acquisition and adoption issues confronting the various federal, state, and local agencies and private organizations (hereinafter called disaster management organizations) that have official responsibility for disaster management. It does not explore the complex issues of IT acquisition or adoption by individuals or private firms for use in disasters; however, it does briefly consider opportunities for leveraging IT systems and services of private-sector firms, citizens, and non-governmental organizations. The chapter starts by considering some of the key barriers to more effective use of IT in disaster management. It then discusses some best practices and design principles that would help address these barriers. It concludes with a discussion of roadmapping as a technique for guiding overall investment in research and development and a discussion of multidisciplinary centers as a way of better coupling IT research and practice.

## OVERVIEW OF NON-TECHNICAL BARRIERS

Many sectors, such as banking, manufacturing, and services, have been able to adopt new IT technologies routinely and aggressively. Some disaster management organizations have also been quite effective in integrating state-of-the-art IT technologies into their day-to-day operations (e.g., the use of Internet Protocol [IP]-based emergency management tools, the use of cell phones to listen in on first responder land mobile radio traffic, and the use of laptops and wireless local area networks). However, in the committee's view, the disaster management community has not been nearly as broadly successful.

The following are among the complicating factors:

- *Disaster management organizations often lack the resources to acquire valuable capabilities.* Responsibility for disaster management is widely distributed among agencies and organizations at all levels of government—with resources and operational responsibilities mainly concentrated at the local level. These organizations have vastly different technologies and capabilities. These characteristics lead to highly scattered adoption and lengthy adoption cycles and a highly fragmented market for disaster management IT. Moreover, many of the organizations are small and have very constrained budgets for IT. Most acquisition resources are focused on capabilities to improve day-to-day operations, whereas disaster management is, by definition, not a routine activity. Some of what agencies do acquire specifically for disaster incidents nonetheless becomes "shelfware"—unused even when the need for which it was acquired arises.
- *Both the development and the deployment of many promising technologies are risky and costly compared with the opportunity presented by the commercial market for these technologies today.* For example, there are sensors that would be very useful for assessing in real time the status of the built environment. However, developing and manufacturing such sensors for the uncertain and highly cost-constrained disaster management market do not constitute attractive commercial opportunity at this time.
- *In most agencies with disaster management responsibilities, there is no one who is charged specifically with tracking IT technology, identifying promising technologies, integrating them into operations, or interacting with IT vendors to make sure that needs are addressed.* Many organizations are too small to grow and support significant in-house expertise, and they naturally look to vendors to provide turn-key solutions, which may mean that the organization's long-term, broad needs are not fully met. Long intervals occur between acquisitions, with the result that any institutional learning that does occur is likely lost in the interim. The acquisition dynamics created by this situation tend to limit the potential market, leading IT vendors to adapt IT technologies only slowly for use in disaster management. There is no focal point for addressing these issues at the federal level, further contributing to the problem. Finally, the complexity of IT systems and the organizational changes that they introduce are often met with resistance and ambivalence by both managers and users, especially in the absence of a technology "champion."
- *Decisions regarding IT tend to be made independently by local organizations that must work together in disasters.* Organizations with disaster management responsibilities are typically highly independent and have lim-

ited regular contact with one another. However, these organizations find themselves having to collaborate in disasters, giving rise to interoperability issues at many levels. State and federal organizations charged with disaster management face similar coordination challenges, further complicating collaboration in responding to a disaster. Acquisition managers concerned about collaboration typically have no place to go to determine if the technologies they are acquiring will interoperate with those of their peers. Further, no mechanism exists for them to synchronize technology acquisitions in order to make them compatible. Recent trends toward the establishment of regional groups to address IT and related disaster management issues are a promising trend.

- *Disaster management is concerned with environments that are intrinsically uncertain and unstable.* This contrasts with the typical IT acquisition environment, where development, deployment, operation, and maintenance take place in fairly well understood and stable environments and where requirements are better understood.

- *Important sources of funds are typically only available once a disaster has been declared and must also be spent in a short window of time.* Funds tend to become available in much greater quantity during a period of time after disaster declarations. Experienced emergency managers are well aware of this recurrent "window of opportunity" effect, and many of them keep IT and communications projects in draft, ready to proceed as soon as a disaster redirects attention and money to their needs. However, these purchases are naturally driven by immediate concerns rather than longer-term considerations.

One conclusion (overly pessimistic in the committee's view) given these barriers would be that advanced IT solutions are impractical for most local governments and emergency management agencies. Such a view assumes that the existing problems are insurmountable, whereas the committee believes that many of these problems can be mitigated if best practices and principles are followed and if appropriate mechanisms are put in place to support their adoption, such as the research centers that couple technology advancement with practice and community-wide technology roadmapping.

Another related potential misreading of the challenge is that technology that is "advanced" or "leading-edge" is necessarily more complex—and is thus unsuited for organizations without considerable in-house technology expertise. In fact, some trends in information technology are in exactly the opposite direction, with advances aimed at reduced complexity from the standpoint of those acquiring, managing, or using the technology. A reflexive avoidance of advanced technology and new developments could thus counterproductively translate into a failure to adopt systems that are more robust, reliable, and usable.

## BEST PRACTICES FOR ACQUISITION

Best practices for acquisition include an emphasis on iterative development; increased opportunities to test and evaluate technology in practice, together with realistic concepts of operations; and design and evaluation processes that allow for strong coupling among practitioners, researchers, and industry.

### From Waterfall Acquisition to Iterative Development

Historically, as in many other areas, the introduction of technology in disaster management has been characterized by a series of major deployments, occurring at intervals sometimes measured in years or even decades. These long cycle times reflect in part the traditional "waterfall" acquisition process. This acquisition model presumes a linear development process that proceeds in stages from development of a comprehensive requirements specification to design, then to implementation followed by integration, next to testing, then to installation, and finally to maintenance. Modified versions of the model acknowledge some role for feedback between each of these stages and preceding ones.[1] They also mirror the typical capital planning cycles of federal, state, and local government and agencies, which have traditionally made periodic, large investments in new systems and capabilities.

Long acquisition cycles are well known to make it hard to incorporate rapid technological change. The doubling of various measures of computing performance every 1 to 2 years places an obvious premium on processes that can more rapidly incorporate new technology. Moreover, this linear process that periodically seeks to produce *the* solution often fails to deliver the expected capabilities. Requirements creep may end up making the ultimate design overly cumbersome, complex, or costly to implement, leading to cost overruns, delays, and even program cancellation. Users, who only have input to the front end of the process, may find that the delivered capabilities do not meet their needs.

Also, new capabilities and technology opportunities that arise after

---

[1]This development process is described in terms of the software development model explicated in considerable detail by Barry W. Boehm, *Software Engineering Economics*, Prentice Hall, Upper Saddle River, N.J., 1981.

the system development leaves the initial requirements stage are difficult and expensive to incorporate. The reason is that many artifacts of a system grow organically. The practical reality is that large systems emerge from incremental additions in ways entirely unanticipated by the designers of the original system. If the original system is successful, users will almost certainly want to add new functionality. The new functionality desired is by definition unanticipated—if the designers had known it would be useful, they would have included it in the first place.

Indeed, it is essentially impossible in practice for even the most operationally experienced IT systems developers to be able to anticipate in detail and in advance all of a system's requirements and specifications. Often users change their minds about the features they want, or (even more difficult to deal with) they want contradictory features. And, of course, it is difficult indeed to anticipate all potential uses. Thus, system requirements and specifications are inherently incomplete, even though they underlie and drive the relationships among various components of the system. Put differently, the paradox is that successful system development requires non-trivial understanding of the entire system in its ultimate form before the system can be successfully developed. System designers need experience to understand the implications of their design choices. But experience can be gained only by making mistakes, learning from them, and having a mechanism to modify and evolve systems overtime as the understanding of both user and designer grows and as requirements and technology evolve.

For these reasons, development methodologies have been developed that presume an iterative approach to building systems. An iterative process uses multiple, short acquisition cycles, which over time deliver and improve on system capabilities. Such a process encourages feedback from users and allows them to play a constructive and central role in a system's evolution. An iterative process requires, among other things, mechanisms for users to provide feedback to technology innovators and providers. (The committee discusses some possible mechanisms for supporting this process later in this chapter.)

With iterative development, systems that initially include limited functionality are often introduced. As users adopt the technology, they have a mechanism for identifying improvements to that functionality and for identifying desirable new features that technology providers can incorporate into the new product versions. The progression of mobile phone functionality to incrementally include increasingly greater performance and a wider range of features is a familiar example of this process.

An iterative acquisition process has other advantages. Often requirements thought to be essential turn out to be relatively unimportant or little used once deployed. The functionality supporting those requirements can be dropped from future product versions, helping minimize complexity creep. Essential features frequently go unidentified until the system begins to be widely used. These features can be added in a more orderly fashion, evolving the system with continuing feedback from users. Incremental introduction of technology also allows one to exploit the current technology "sweet spot"—where the costs of components such as microprocessors are lowest—keeping down costs and making more frequent acquisition cycles possible.

In disaster management a tension inevitably arises between a natural desire to fully meet demanding or perceived unique requirements and the cost and speed of development and deployment. Disaster management professionals often say that they must be able to depend "absolutely" on the technology they employ—noting the life-or-death nature of their work. An iterative process allows time for users to build trust in the system's ability to deliver on those critical requirements and a mechanism for providing feedback to request (or demand) changes as needed. It also allows an opportunity to minimize initial demands for unique requirements involving specialized equipment and maximizing the opportunity to incorporate "commodity" components, thus minimizing cost and delays.

## Metrics-Driven Investment

As the saying goes, one can only manage what one can measure. The resources available for disaster management are limited, and decision making always involves tradeoffs. To motivate the IT expenditures needed to provide adequately for disaster management, there must be an understanding of the benefits that are obtainable. Weighing the available benefits from particular IT investments against the returns on other sorts of investment is challenging. When considering the effects of disasters, these tradeoffs can easily be driven by emotions, even more than in many other sectors. Having metrics allows an analytical assessment to be made, comparing the costs of preventive and mitigating investments with the likely impacts of disasters, and with other potential investments. The sections that follow briefly discuss several aspects of metrics-based decision making. The development of suitable metrics to guide investment in IT for disaster management is a topic for further research and something that a roadmapping effort (described later in this chapter) might address.

### Estimating Risks

Estimating the risks of infrequent events is hard, but failing to consider risks explicitly cripples any rational decision-making process. Gathering the necessary information will necessarily be an iterative process, with initial information providing a basis for further discussion, expansion, and revision. An additional benefit of systematizing this process is the potentially useful feedback on needs and opportunities that it can provide to the technology research and development community.

## Costs and Benefits

The economic model needed to assess the trade-off of the costs versus the benefits of investing in technology for disaster mitigation differs from business investment models. Typical business investments are related to a steady income stream, not to a variety of infrequent future costs and benefits. An economic model for disaster management must combine initial investments, ongoing costs, and infrequent events. An investment is based on the net present value, computed using some discount rate for all those components. Those rates incorporate the expected lifetime of the assets and the risks associated with deriving income from those assets over that period. For business IT investments, those rates typically vary from 12 to 20 percent. Communication infrastructure has used much lower discount rates in the past, but the merger of those technologies is forcing the rates for those investments upward.

Any technology deployment has initial costs, as well as ongoing maintenance and training costs, and a finite life. This long horizon requires using discounting of future benefits, as well as ongoing costs. Since the occurrence, magnitude, and timing of future disasters are uncertain, appropriate discount rates may have to be quite high, so that results adequately reflect the intuition of the participants and funders of disaster mitigation expenses. While economic cost estimates of disaster impacts are never precise, they do provide order-of-magnitude estimates needed to allow projects and proposals to be ranked.

Estimation of the savings resulting from reduced impact due to mitigation efforts is particularly difficult. Many IT benefits will be due to being able to respond more rapidly. Developing models on how faster response can reduce eventual costs is a substantial, but interesting, challenging, and rewarding task. An actual economic quantification of the cost of disaster mitigation versus the benefits obtained could be a fruitful area for research.

## Use of a Cost-Benefit Model

Any recommendation for new and increased outlays must be accompanied by a quantification of their benefits. While costs are easy to quantify, the benefits of disaster mitigation are hard to quantify, but a reasonable attempt is required. It is expected that the costs of improving the technology available for disaster mitigation will be offset by substantial benefits accruing to the country. The most important of these benefits cannot be directly quantified, since they represent the human dimension: reduction of suffering, preservation of family stability, and prevention of losses of items of purely personal value. Other benefits of disaster mitigation can and should be quantified.

The low frequency of major disasters greatly reduces the priorities that local planners, faced with many short-term needs, actually assign to accumulating and maintaining the resources that are adequate for dealing with disasters. While some supplies can be stockpiled for decades, IT becomes obsolescent much faster and requires an ongoing infusion of funds. In the commercial world, a spending rate of 15 percent of the initial and upgrade investments is expected. Costs are reduced when obsolete systems are taken out of service. For many systems, the military tends to spend less annually but is then faced with huge, wholesale replacement costs every 12 years and has an inadequate system for more than half of that period.

Readiness for mitigating disasters requires a modest but steady investment in technology. The total benefits are due to cost reduction that occur at unpredictable times and are of unpredictable magnitude. Major, quantifiable benefits are due to infrequent events. For an individual county, the investment in disaster mitigation technologies appears to represent an instantaneous expense, sometimes aided by state or federal grants. Maintenance costs are in different budgets and are hard to assess.

## Technology Evaluation

The decentralized nature of disaster management, spread across thousands of agencies—from the smallest volunteer fire department, to sophisticated urban police departments, to state, regional, and federal agencies—presents particular problems for effective technology evaluation and diffusion. Today, many managers responsible for the acquisition of technology for public safety and emergency management are, quite understandably, unable to keep-up-to date with the volume of technology and choices available. Managers often rely on vendors to tell them what they need and have to base decisions largely on the often-conflicting "advice" of various vendors.

Professional conferences, workshops, and other meetings held by public safety and emergency management associations are one mechanism for facilitating diffusion of the latest technology. IT capabilities, most notably the Internet, may prove useful as well by providing a conduit for sharing and discussing information about what works.

More systematic approaches to evaluation would likely yield deeper and broader technology adoption over the long term. One option is to make use of formal mechanisms for providing unbiased evaluations and guidance—a sort of *Consumer Reports* for disaster managers. The military's experience with technology demonstrations, described in the next section, may provide one model for this type of "clearinghouse" approach. Important differences exist between the defense and disaster management contexts regarding technology evaluation. For instance, the military has a somewhat-well-defined acquisition chain that flows from initial ideas to deployment agencies. In civilian disaster management, development and procurement are far more decentralized. Decentralization introduces hand-off issues for successfully demonstrated technology. Still, adapting the lessons of military technology transfer to civilian disaster management yields at least two insights.

The first insight is that technology demonstration will be successful to the extent that more knowledgeable technology adopters are available for experimentation. The next two sections—on processes that bring technology developers and practitioners together and on building capacity at the intersection of IT and disaster management—discuss mechanisms and examples for growing the capacity of practitioners as knowledgeable technology adopters.

A second insight is the importance of an honest broker serving as a neutral technology clearinghouse that can help provide the expertise to identify and evaluate technology. There are a few examples, from which others have successfully learned, where local and state agencies have taken the lead in demonstrating the viability of a technology. Some states (e.g., South Dakota and Indiana[2]) have taken on this role, identifying and evaluating technology, infrastructure, and services, and in several cases providing one or more of these to local agencies.

Private-sector integration centers aimed at bringing together diverse technologies can have value in getting vendors to make their products work with those of other vendors. But, they are necessarily designed to promote both their particular partner's products and their own consulting and integration services. These vendor-driven efforts will likely fall short of being truly neutral.

One option for achieving the neutrality of an honest broker to vet technology intended for disaster management is used for similar reasons in other government mission areas. It is the Federally Funded Research

---

[2]Based on the testimony of Otto Doll, Bureau of Information Technology, state of South Dakota, and Dave Smith, Indiana Integrated Public Safety Commission, at the committee's June 2005 workshop.

and Development Center (FFRDC) model. FFRDCs are independent, nonprofit entities sponsored and funded by the U.S. government to meet specific long-term technical needs. FFRDCs typically assist government agencies with scientific research and analysis, systems development, and systems acquisition. They draw together expertise and perspectives from government, industry, and academia to address complex technical issues.

An FFRDC for disaster management would not have any operational responsibilities. Rather, it would serve the disaster management community by identifying, developing, and assessing technologies and concepts of operation for using those technologies.

## Processes That Bring Technology Developers and Practitioners Together

An iterative development process goes hand in hand with an acquisition process which assumes that technologies and organizational processes will co-evolve. Coordinating technological advances and organizational process changes requires new knowledge and skills on the part of both practitioners and technology developers and new relationships between them. Such coordination depends on practitioners and developers gaining a better understanding of one another's methods and on mechanisms that maintain dialogue between them in order to identify promising technologies, define appropriate uses for them, and evaluate and disseminate the outcomes.

The Department of Defense has a broad set of programs aimed at bringing together technology developers and users, speeding innovation, and transitioning it into use in the field. One notable model with considerable applicability to disaster management is the Advanced Concept Technology Demonstration (ACTD) model,[3] which also pays particular attention to the interplay between technology and organization. An ACTD is used at the phase where promising technologies have been developed together with a vision of how they could be used. An ACTD provides a framework in which to assemble a group that is willing to be an early adopter and a context into which the technology can be inserted and evaluated. A well-run ACTD includes a phase where the system, organization, and technology are all analyzed together, and modifications to each are identified and implemented.

A particular strength of the ACTD approach is that it recognizes that

---

[3]A description of the goals of an ACTD is available at http://www.acq.osd.mil/actd/transit.htm.

it is not enough to build the technology. One needs to analyze the organization and look at how processes are going to change as a result of having the technology. The idea is to simultaneously develop requirements for organizational change and adaptations of technology to fit that.

To use the ACTD approach effectively, patience is required. Otherwise, the design phase may be overly compressed to the detriment of the ultimate product. Promising innovations may wither while waiting to be adopted, or problems identified during the acquisition phase may result in a technology's being abandoned before researchers are able to find solutions.

### Building Capacity at the Intersection of IT and Disaster Management

The committee heard from state and local agencies that one of the major barriers to advancing practice and adoption of technology was a lack of resources to allow staff time for ongoing development of technical expertise. Yet, without the in-house development of technology expertise able to draw on external resources (such as centers of excellence), adoption of technology will continue to lag and is unlikely to be optimally implemented when adopted.

The interdependence of technology and practice means that developing a cadre of experts at the intersection of disaster management and IT is likely to yield significant payoffs. Expanding the human assets available involves both the promotion of cross-fertilization between the technology and practitioner communities and the promotion of a culture of innovation in both. Such a cadre of people will be more astute at translating user requirements to technical need, and will serve as a self-reinforcing feedback mechanism between technology advances and disaster management practices.

A number of mechanisms could contribute to increasing human capital along these lines. These include both mechanisms for fostering innovative environments wherever possible and mechanisms for disseminating their results elsewhere. For example, programs could be established to support fellowships, field tests and other experiments, and training and educational activities. Also, programs that incorporate both disaster and IT expertise could be funded to analyze the performance of systems after a disaster.

Federal grants could support creation of expertise within state and local agencies by, for instance, sending people from public safety agencies to regional centers for training and to interact with technology experts and other practitioners to stay abreast of the latest developments in both practice and technology.

## Exploiting Open, Practitioner-Driven Processes

The rapid pace of technological change, the growing complexity of technology, and perceived economies of scale are driving many disaster managers to move from being owner-operators of their IT and communications to being customers of contractors and service providers. Even where public safety and emergency management agencies retain ownership of their IT and communications assets, they increasingly have become reliant on vendors for information on what is technically possible and worthwhile.

Formal acquisition and management processes are generally designed to enforce appropriate procurement standards. However, they also run the risk of filtering out many useful, original ideas and possible contributions from outsiders and can present insurmountable obstacles to the acceptance of new technologies, unless and until they have been converted into profitable commercial products. Even when emerging capabilities are tracked and assessed, traditional design and acquisition methods are ill-suited to keep pace with accelerating shifts in technology.[4] As discussed below, there are a number of opportunities to broaden the set of users that participate in the shaping, development, and evolution of IT systems.

The recent movement of open source software and standards development is one recent expression of this voluntary impulse. (See Box 3.1 for background on open source software and open standards.)

Although some of the resistance to the use of open source software in disaster management might ultimately be traceable to active opposition by commercial vendors, there are also structural obstacles to non-commercial innovation within disaster agencies. This is regrettable, since many technical innovations begin as non-commercial experiments, though only a relatively few survive the road to commercialization. Despite this resistance, there are some examples of how the open source software/open standard development model has been applied to disaster management. (See Box 3.2 for one such example.)

Several strategies are available for extracting the most valuable items from the non-commercial offerings of the volunteer and open source communities:

- Organized experimentation and evaluation of non-commercial and pre-commercial technology. (The city of San Diego and the state of Cali-

---

[4]J.C. Herz, Mark Lucas, and John Scott, "Open Technology Development, Roadmap Plan," Version 3.1 (Final), Department of Defense, Washington, D.C., 2006.

> **BOX 3.1**
> **Open Source Software and Open Standards**
>
> Open source software and open standards are two aspects of a cooperative approach to information technology that has deep roots in the Internet. Open source refers to the internal programming—the source code—of an application. In traditional proprietary software, the secrecy of the source code is the foundation of its commercial value. Access to source code is restricted, and details of its operation are disclosed only in general terms. Open source software, on the other hand, is freely published and generally visible. While some financial value can be recovered in fees for ancillary services (e.g., consulting and sales of reference books), the chief reward for the creation of open source software accrues to the reputation of the programmer or programmers who contribute to its creation.
>
> Although much open source software is the work of individual programmers or small, discrete teams, the "reputation economy" of open source software development lends itself to the formation of large ad hoc collaborations, often involving developers widely distributed geographically and interacting via the Internet, competing for recognition of the quality of their contributions instead of for commercial equity.
>
> The open source approach has produced some of the world's most popular software, such as the widely used Apache Web server program and the Linux computer operating system. Proponents of open source software argue that it tends to be more secure (since security weaknesses cannot be hidden within proprietary programming) and more flexible (since it can be readily customized) at lower costs than commercial software products. They suggest that the lack of financial constraints leads to software that is "problem-oriented rather than profit-oriented" and that the lack of a commercial incentive to lock in customers to a particular program helps preserve market efficiencies over time.
>
> Critics of open source software dispute the cost-effectiveness of such "free" software, countering that it demands more of IT staff and cannot be proven to be

fornia, among others, have ongoing programs under which non-commercial innovators can test and demonstrate their ideas in parallel with disaster exercises.)

- Assistance in refining useful non-commercial technologies. Such assistance might be provided in a number of forms, including coordinating small-business loans and developmental microgrants, brokering introductions to commercial implementers, and offering user feedback and review to non-commercial development projects.
- Use of selected non-commercial advisers as independent sources of information and reviewing IT and communication plans under the auspices of an advisory organization.

Judged on the merits of their results, open source software/open standards developers could become valuable members of the disaster

> any less expensive to maintain than commercial software (for which 15 percent of the purchase price annually is a common estimate of maintenance costs). They also warn that the design of open source software may reflect the interests of the developers more than the needs of the users.
>
> Open standards, by contrast, are blind to the internal composition or economics of any particular program or device. Open standards specify certain external behaviors of a program or device in order to ensure interoperability among various implementations. Open standards may describe file formats, communication protocols, or equipment configurations. They are deemed "open" to the extent that they are published so that products and applications can implement them without licensing or other costs.
>
> Arguably the most successful open standards effort to date has been the Internet itself. By publishing a collection of non-proprietary technical standards (the Internet protocols) for unencumbered use, the creators of the Internet enabled data exchange among diverse computer systems and software packages. As a result, the data and functionality available on any Internet-connected computer now vastly exceed the usefulness of that same computer standing alone.
>
> The open standards movement and the open source community share an emphasis on collaboration and cooperation as sources of value. This "network economics" view sees value as the product of connectivity among numerous entities, as opposed to the more traditional view that value is a function of scarcity. (The classic illustration of the former line of thought is that the value of the first fax machine was nothing until there was a second one. And the more fax machines there were in the world, the more valuable each of them became.)
>
> Open source software and open standards are not so much challenges to traditional commercial approaches as they are extensions and supplements that take advantage of voluntary collaboration and cooperation, especially in low-frequency, high-risk, and high-uncertainty applications where commercial incentives alone may not yield needed results.

management community. Successful examples of applying this model in other areas range from widely used programming languages and operating systems to the Internet. They have a few key common elements important to their success:

- Commitment to open source and open standards processes, with supported organizations for managing them, to harness the energy and skills of individuals and small groups that best understand the needs of the community;
- Small but very-high-leverage investments in accelerating technology dissemination by funding organizations to provide "supported shareware"—reference implementations of technology robust enough for users to use and evaluate, while open to all members of the community (users and technologists alike) to refine and improve and extend to others. This is a highly proven model for transitioning to robust, commer-

> **BOX 3.2**
> **Common Alerting Protocol (CAP)—An Example of Open Standards Development Applied to Disaster Management**
>
> One recent example of open-standards-style development in support of disaster management is the Common Alerting Protocol (CAP). This international data standard for alerting and warning messages was initially developed by an ad hoc group and subsequently attracted the attention of two non-profit organizations (the ComCARE Alliance and the Partnership for Public Warning). Later, the CAP standard was adopted by the international Organization for the Advancement of Structured Information Standards (OASIS) standards body and by a number of federal agencies (including Department of Homeland Security, Department of Defense, National Oceanic and Atmospheric Administration, and U.S. Geological Survey), and state and local organizations began to use CAP in a variety of emergency alerting and notification applications.[1] Ultimately, the international OASIS standards group partnered with U.S. and international disaster response programs to formalize CAP and an associated family of Emergency Data eXchange Language (EDXL) standards[2] for disaster management.
>
> ---
> [1]See Partnership for Public Warning; available at http://www.partnershipforpublicwarning.org/ppw/cap.html. See also *Common Alerting Protocol, v. 1.0*, OASIS Standard 200402, March 2004; available at http://www.oasis-open.org/committees/download.php/6334/oasis-200402-cap-core-1.0.pdf.
> [2]See *Emergency Data Exchange Language (EDXL) Distribution Element, v. 1.0*, OASIS Standard EDXL-DE v1.0, May 1, 2006; available at http://www.oasis-open.org/committees/download.php/18774/EDXL-DE%201.0%20Standard.html.

cially supported products without locking government to limitations resulting from single sources of supply;[5] and

• Support for the development and exchange of information about best practices through Internet support for forums of information dissemination and discussion, for example via blogs and wikis.

Full exploitation of the open source software/open standards development model requires organizations and institutions that take on new responsibilities to guide and carry out activities. Many different types of

---
[5]The World Wide Web took off in part because the National Center for Supercomputing Applications at the University of Illinois was funded to provide a supported shareware Web browser, Mosaic, which went on to form the underlying code basis of every commercial Web browser in the world.

organizations could fill this role. However, there is a tremendous community-building opportunity if the role is filled by an organization that has an educational component, has systems development and support skills, and is rooted in the practitioner communities. The ideal structure would be one where training of both IT developers and technology-savvy practioners is performed in an environment that includes testbed technology innovations and where the practitioner community and IT development community could learn from each other, providing a positive feedback loop for the development of IT solutions grounded in practical realities.

## TRAINING AND THE IMPORTANCE OF ROUTINE USE

Disaster managers emphasize the importance of technology being used on a routine basis by practitioners if it is to be used effectively, or at all, during a disaster event. Understandably, practitioners turn to the things they trust and are most familiar with, especially in the high-stress situation of disaster response. Training is important, but it often occurs only once, shortly after a technology is introduced. Technology that sits idle until the onset of an event will likely remain unused. The committee heard of instances where responders did not know or remember where the technology was or how to access it when it was eventually needed. Frequent training is also necessary because the technology itself often changes rapidly.

In each of the areas of disaster management activity, most of the organizations and personnel (e.g., law enforcement, firefighting, emergency medical treatment, sea patrol and rescue) involved have regular day-to-day responsibilities and activities different in kind and in magnitude from major disaster operations. Training and equipment provided for disaster responsibilities may not be optimum for day-to-day activities, so that compromises or separate parallel training and equipment may be necessary. Full consideration of cost, complexity, and other issues associated with this problem are such that implementation is difficult.

Training programs are expensive and take critical resources out of service for the duration of the training. It is often logistically difficult to arrange training for huge public safety organizations, with training sometimes spread over many months. The less relevant the lessons are to their daily tasks, the less likely practitioners are to retain lessons from the training. Yet, IT developed for disaster management is frequently designed for use only during major events, and using it effectively requires special knowledge or familiarity based on experience (which rarely exists). The result is that much IT goes unused or underused when an event occurs. Building adaptable tools that people use every day is critical to those tools being used in a disaster.[6] Indeed, routine use is more important than specialized training in building the competence and confidence required to use a technological capability successfully, especially in the high-stress situation of disasters.

One partial solution is to seek technology that has the broadest possible application in daily operations, with smooth transition (from a user's perspective) to functionality required only in handling disaster events. The switch over to disaster operations should be as seamless as possible. As an example of what to avoid, the committee saw an operations center during a site visit where two separate software applications with overlapping functionality coexisted; one application was used during routine operations, the other only in disaster situations. The application used on a daily basis was widely and effectively used. The application with additional specialized disaster functionality was ill understood by all but a few personnel.

## LEVERAGING "NON-OFFICIAL" INFORMATION TECHNOLOGY

Although disaster management is typically associated with government response agencies, the true boundaries of the disaster management community are difficult to draw. Non-governmental organizations of various kinds, government agencies not normally concerned with disasters or public safety, trained volunteers and emergent ones (especially within the victim population itself)—all these are active participants in disaster management. Private voluntary organizations play important roles in disaster response and recovery activities, including the distribution of food, water, and other supplies and the provision of shelter. Amateur radio operators have long played a role in disaster communications and have been formally incorporated into disaster communications planning and procedures. Moreover, in a disaster, many individuals and organizations will step forward as volunteers. Some will be those who happen to be present at the scene of a disaster. (For example, a number of live rescues after an earthquake are performed by bystanders, not official responders.)

Many volunteers prove to be crucial contributors to a disaster response and possess resources or expertise not otherwise available. Others

---

[6]This is one of the conclusions of Sharon Dawes, Thomas Birkland, Giri Kumar Tayi, and Carrie A. Schneider, *Information, Technology, and Coordination: Lessons from the World Trade Center Response*, Center for Technology in Government, University at Albany, State University of New York, June 2004; available at http://www.ctg.albany.edu/publications/reports/wtc_lessons/wtc_lessons.pdf.

may consume more of official responders' time than their contributions are worth. Some create confusion for the response system: Are they responders or victims? Are their motives altruistic, selfish, or even criminal? Are they actually who they say they are? Nor are the priorities of volunteers necessarily well aligned with those of officials. As a result, volunteers and their contributions to official agencies are generally filtered through coordinators and programs that act to limit their impact on the agencies' official programs.

Recent IT advances offer new ways for volunteers to contribute to disaster response and recovery efforts. For example, the following are just a few of the many volunteer activities in the response to Hurricane Katrina:

- *Web-based information aggregation systems* at http://katrinalist.net and http://www.disastersearch.org. These are systems for posting and aggregating information about people affected by Hurricane Katrina and resources available for those directly affected by the disaster.
- *An online repository* at http://www.hurricanearchive.org of user-contributed information about the hurricane experience. This repository has a "map browser" page that incorporates the use of Google maps. A similar map display was used on another site during the immediate aftermath of Hurricane Katrina to provide information about damage at specific sites throughout the city of New Orleans.
- *A toll-free locator service* established by MCI in the aftermath of Hurricane Katrina that provided toll-free numbers where volunteers could register themselves (877-HELP-KAT) and could search for missing family and friends (866-601-FIND).

Exceptional circumstances require the rapid construction of working relationships among groups and individuals with very different backgrounds and priorities. Network technology has created new opportunities, as the above examples demonstrate, for rapid establishment of working relationships, but much remains to be learned about the practical management of these crucial "adhocracies." The large number of private companies providing IT-related services, wide access to information systems, the natural support that the Internet provides for "virtual communities" of users and developers, and the relative ease with which new methods of communications can be tried out have led to the creation of many novel communications mechanisms that have applications for disaster management. The success of technologies like Google's search engine, Wikipedia, blogs, and Google maps is evidence that information technologies that provide user-driven information access are powerful enablers.

These tools are revolutionary in that the information providers are not forced into pre-authorized pre-formatted reporting and because users are presented with multiple sources providing data on any query. At the same time, users are forced to sift, evaluate, and decide what information is helpful. And old problems, such as conflicting and redundant efforts, unclear motivations, lack of accountability, inaccurate information, and differing priorities are all present.

Problems or not, it can be safely assumed that future disasters will bring about the spontaneous creation of similar non-official uses of IT—and most likely completely novel ones as well. The obvious potential value of these initiatives to various aspects of disaster response and recovery activities coupled with the potential for problems should drive disaster managers to seek constructive ways to incorporate them into disaster management practice whenever feasible and appropriate. Careful study should be undertaken with the goal of making iterative improvements in leveraging these non-official, emergent activities so that they eventually become part of the standard inventory of techniques and procedures used to deal with disasters.

## EFFECTIVELY USING COMMERCIAL COMMUNICATCIONS SERVICES

Commercial communications services offer potentially large cost savings over dedicated systems but raise questions as to whether stringent capacity and reliability requirements can be met. Striking the right balance in the face of market forces that drive commercial service providers will require careful and ongoing evaluation by public safety agencies.

Public safety agencies maintain extensive infrastructure independent of the commercial infrastructure available to the general public. There are a number of important reasons for this. The National Task Force on Interoperability (NTFI) guide for public safety officials "Why Can't We Talk?" asks the question, "Why can't they just use cell phones?"[7] It answers the question by noting that responding to disaster situations, where every second counts, requires reliable, dedicated equipment. Public safety officials cannot depend on commercial systems that can be overloaded and unavailable; experience has shown that these systems are often the most unreliable during critical incidents when public demand over-

---

[7]National Task Force on Interoperability, "Why Can't We Talk?: Working Together to Bridge the Communications Gap to Save Lives: A Guide for Public Officials," February 2003, p. 11; available at http://www.ncjrs.gov/pdffiles1/nij/204348.pdf.

whelms the systems. The NTFI guide summarizes the "unique and demanding communications requirements" for optimal public safety communication systems as including dedicated channels with priority access, reliable one-to-many broadcast capability, highly reliable and redundant networks, the best possible coverage within a geographic area, and equipment designed for quick response—in short, something officials can control and count on.

And yet, the use of commercial infrastructure by first responders and public safety officials is increasing. This is happening for several reasons. First, commercial infrastructure is much less expensive than dedicated, purpose-built public safety communications infrastructure. Second, it is often what is at hand during an incident. Third, it helps achieve resilience by provided redundancy and diversity, potentially at relatively low costs. (See the detailed discussion on redundancy and diversity to achieve resilience later in this chapter.) Efforts, both formal and informal, are under way to make additional use of commercial services. Commercial infrastructure used today for disaster management includes the following:

- Cellular voice networks,
- Cellular data networks for mobile data terminals (such as those in police cars and ambulances),
- Cellular data networks for text messaging and e-mail services,
- Satellite phone networks,
- Ad hoc wireless networks,
- Municipal wireless networks (some of these are dedicated networks for public safety personnel),
- Traffic management systems,
- Aerial photo systems,
- Radio over IP systems,
- Cable/public television broadcasting systems,
- Video monitoring and conferencing systems,
- Paging systems,
- Telephone bridges,
- Internet services, including (but not limited to) Voice over IP,
- Remote hosted incident management services (e.g., eTeam, WebEOC), and
- Commercial imagery satellites and image databases.

Commercial hardware, software, and network services are subject to intense competitive price pressure. Maintaining profitability entails a vigorous effort by the providers to minimize costs. But many of the steps available to control costs—reduction of component count, use of the least expensive components available, avoidance of unused "overhead" capacity, and the like—can directly reduce the resilience of the product when it is used in unusual circumstances.

For example, unused capacity is expensive for shared telecommunications systems such as the public telephone network, cellular phones, and satellite systems. System operators conduct careful statistical studies to estimate the level of capacity needed to satisfy most of their customers under most circumstances. Any capacity in excess of that is, from a business perspective, wasted, and operators have a fiduciary duty to avoid such waste. Unfortunately, this means that during exceptional surges in traffic, as occur during disasters and even on holidays and during certain sporting and cultural events, the available network capacity is exceeded and service becomes unreliable. (This economic phenomenon is sometimes obscured by memories of successful use during previous disasters, when the system in question was new and not yet fully loaded to profitability.)

## DESIGN PRINCIPLES

A high degree of reliance on turn-key systems has meant that disaster management organizations have not had to pay a great deal of attention to the underlying design issues that ultimately affect the functionality of their IT systems. Today, pervasive networking and the increasing importance of off-the-shelf technologies make it possible for organizations to assemble a greater portion of their disaster management systems from components provided by different vendors. Doing so successfully will require that those organizations take on more responsibility for the design of those systems. In the course of its work, the committee has identified four system design principles that have particular importance for disaster management systems:

- *Build emergency management systems for effective scaling from routine to disaster operation.* System designs that feature wrenching and risk-laden transitions between daily operational mode and "disaster" mode can make bad situations worse. All system components should be used regularly, even if not at full scale, by the people who will need them in a disaster. Familiarity smoothes transitions from routine to disaster operations. Indeed, people tend to avoid or are ineffective with the unfamiliar.
- *Exploit redundancy and diversity to achieve resilience.* Improving overall resilience can be achieved using many different techniques in combination. One approach is to "harden" systems, designing them to higher operational standards than those typically applied to commercial services and technology. A valuable alternative, which offers both the promise of

improved performance and lower cost, is to exploit redundancy and diversity to improve the overall resilience of an IT system.

- *Design systems with flexibility, composability, and interoperability as core guiding principles.* Each of these characteristics has important, wide-ranging implications for system design. Systems that place a premium on flexibility and agility lend themselves to the sorts of ad hoc use that often is needed in a disaster. Systems composed of standard components will improve the ability to evolve them incrementally and reduce reliance on any single vendor. Systems with designed-in interoperability points greatly facilitate interoperation with other systems.
- *Distinguish between the user interface and the underlying technologies used to deliver a capability.* A particular user experience and the particular technologies used to deliver it need not be the same. Consider, for example, that first responder radios encapsulate multiple attributes—push-to-talk, the form factor of the handset and microphone, and push-to-talk communications within defined groups—that could be unbundled and repackaged for more effective use. Similarly, future handheld communications devices might employ cell phone technology (not commercial cell phones themselves) but with appropriate adaptations (in durability, form factor, or frequency) that meet disaster management requirements yet leverage lower-cost commercial technology.

Each of these principles is discussed in more detail below.

### Effective Scaling from Routine to Disaster Operation

Disasters are sufficiently different from day-to-day life events to be perceived as a completely separate mode of operations involving different organizational structures and priorities, different attitudes and skills, different methods, and different technologies. This impression is reinforced by a collection of binary words and concepts: disasters are "declared," emergency plans and facilities are "activated," resources are "mobilized" and then later "demobilized."

Yet this clear-cut distinction between "disaster mode" and normal operations is in some sense a matter of administrative convenience. Actual disasters vary enormously both in scale and quality. Some, such as earthquakes, terrorist explosions, and electrical blackouts, come on suddenly and with a minimum of warning. Others, such as riverine floods, hurricanes, and food shortages, come on relatively gradually. Some, such as infectious disease epidemics and bioterrorism attacks, might actually be in progress for some time before being recognized. By the same token, some disasters conclude quickly, allowing a rapid shift from response into recovery, while others persist for longer and sometimes unpredictable periods. And while the generalized incident command functions of command, operations, plans, logistics, and finance are generic and applicable to all hazards, the specific makeup and activities of each function can vary considerably depending on the nature, scope, and phase of the particular event.

In addition to the challenge of matching the response to the situation, these broad ranges of possibilities create challenges for managing the transition into and back out of "disaster mode." Many problems in disaster response may be artifacts of these transition challenges, especially as regards interpersonal and interorganizational communication and coordination. Differences in situational awareness and assessment and resulting differences in the timing and degree of "activation" between organizations and individuals can lead to procedural and perceptual disconnects and to falsely calibrated expectations. For example, in 1991 California regional emergency management personnel near a major urban wildfire experienced difficulty in mobilizing their headquarters to respond to resource requests. The more distant headquarters had not yet received tactical and media reports of the extent and rapid growth of the blaze and thus had not yet fully activated.

This tendency toward representation errors and "mode binding" may be exacerbated by the tendency to focus on worst-case scenarios in planning and exercises. The tacit assumption that a worst-case scenario will exercise all emergency management functions overlooks the problems inherent in transitions and the ambiguities of borderline situations that may or may not turn out to be disasters or that are evaluated differently by different response entities.

Thoughtful use of IT could do a great deal to mitigate the problem of transitions, largely by reframing occasional disaster activities as continual processes that can be scaled up or down without creating discontinuities. At least four potential benefits could be realized.

First, the interconnection of emergency management applications using shared data standards could permit activities such as situation reporting and resource management to be turned into continual "flow" processes rather than "batch" activities based on paper-based document-oriented systems. This conversion could reduce the time delays and the unintended synchronization of reports, requests, and orders inherent in fixed reporting cycles. It could also permit more rapid and precise detection of trends and conditions by allowing observations to be shared at a rate suited to the phenomena observed rather than one dictated by a fixed reporting schedule.

Second, a continually updated shared knowledge base, in the form of a "common operating picture," could reduce various difficulties that can result from different entities having different editions of situation and resource reports at a single point in time.

Third, the greater "virtualization" of emergency management processes using network technologies could reduce the impact of disasters on day-to-day operations by allowing many organizations and individuals to participate in the emergency information system from their regular workplaces. It also allows for more phased relocation to emergency facilities, thus reducing the vulnerability of the emergency management system that can result when all participants are in transit at the same time.

Finally, in addition to reducing the necessary size of physical emergency operation centers, greater virtualization might also permit a more gradual activation of emergency procedures, one that can be more closely matched to the particular "curve" of a particular disaster. This could reduce both the disruption and missteps associated with all-at-once activations and the tardy activations that sometimes result from hesitance to take the disruptive and expensive step of dislocating large numbers of key personnel.

One key to obtaining each of these benefits would be to abandon contingent "in case of disaster, break glass" systems and procedures in favor of an "always on" approach enabled by modern IT and communications. It is also critical to understand the degree to which IT can effectively complement, support, and substitute for face-to-face communications.

## Redundancy and Diversity to Achieve Resilience

One of the fundamental dynamics of disaster management is the tradeoff between efficiency and resilience. At a technical level, communication can be disrupted by many factors, such as the following: physical destruction of equipment or infrastructure (e.g., towers, cables, and substations), loss of power, interoperability problems, and environmental factors (e.g., obstacles for wireless). *Hardening* devices and infrastructure is an obvious way to reduce failures of equipment and infrastructure and reduce the chances of power loss. While it is not economically feasible to harden all critical equipment, improvements are certainly possible. For example, it should be possible to develop tools to analyze the robustness of the communication infrastructure in certain disaster scenarios and to apply resources optimally to harden those at greatest risk.

While hardening equipment and infrastructure continues to offer potential for improvements, other means of achieving resilience in IT systems for disaster management are often discounted or overlooked. *Redundancy* and *diversity* are two well-known techniques for building resilient (reliable, robust) IT systems.

Redundancy can be provided inside a network that uses a specific technology—for example, by building additional relay stations or access points into the infrastructure. This would make it possible to route traffic over alternate paths if some components fail. The ability to establish parallel redundant networks quickly can reduce down time. Two examples are (1) parallel wireless networks based on commercial technology and (2) cellular-on-wheels (COW)—movable cellular sites with satellite backhaul to establish cellular bubbles using dedicated spectrum.[8] It is necessary to plan for such deployments in advance. Equipment must be available, spectrum must be allocated, and people must be trained. Also, organizational support must be created (e.g., directory services).

Technological diversity in the basic infrastructure for all organizations (including large and small businesses, government agencies, and community organizations) would reduce the impact of disasters.[9] A related risk is that a technology monoculture, in which a technical attack in a single operating system or other piece of software could cripple response activities across the board. Indeed, integrating disaster planning efforts across private and public IT systems that embrace diversity and redundancy could lead to significant improvements in the overall reliability of information and communications capabilities in the event of a disaster.

As discussed earlier in this chapter, one important opportunity for achieving better resilience is by using commercial technology and services. Commercial technologies offer significant opportunities for both better redundancy and increased diversity. Two examples of this opportunity are cellular and wireless networking technologies. Nearly everyone uses cell phones on a daily basis, so they will naturally continue to use them during a disaster. Besides traditional voice calls, cell phones also increasingly support other communications capabilities—for example, push-to-talk, text messaging, Web access, and instant messaging—all of which may be useful and familiar. Wireless networking is also becoming ubiquitous in populated areas and is supported on multiple mobile de-

---

[8]COWs have been deployed for wilderness firefighting, and Qualcomm deployed an improvised system that included movable switches (switches-on-wheels, or SOWs) with satellite backhaul to create cellular bubbles in the aftermath of Hurricane Katrina.

[9]Sharon Dawes, Thomas Birkland, Giri Kumar Tayi, and Carrie A. Schneider, *Information, Technology, and Coordination: Lessons from the World Trade Center Response*, Center for Technology in Government, University at Albany, State University of New York, June 2004; available at http://www.ctg.albany.edu/publications/reports/wtc_lessons/wtc_lessons.pdf.

vices such as laptops and handheld devices. These technologies are commercially supported and widely used, and the equipment is relatively inexpensive. These characteristics naturally lead to IT capabilities that are both redundant and diverse.

### Flexibility, Composability, and Interoperability as Core Guiding Principles

Systems designed to exhibit the characteristics of flexibility, composability, and interoperability as core guiding principles can produce many long-term improvements with wide-ranging implications for the effectiveness of IT use for disaster management. Each of these characteristics has important implications for system design.

Almost by definition, one cannot anticipate every contingency that must be dealt with in a disaster. In disaster management, where information channels can be interrupted and informal channels established without official sanction, developing and encouraging these characteristics are especially important, though often counter to the urge for control.

This observation suggests favoring composed systems consisting of diverse components over turn-key, integrated solutions. Composing systems from diverse (standard) components allows selection of best-of-breed solutions, enables federated solutions made up of diverse technologies with carefully chosen operational boundaries at which interoperability can be managed, and provides greater overall robustness through diversity.

Another important design principle is to build systems assuming future needs for interoperation with other systems. By constructing systems using layered architectures with potential points of interoperation designed in from the beginning (e.g., by designing to established standards), systems can be made interoperable without requiring technological monocultures.

Composability offers further advantages. The short-term economies attributed to lockstep standardization can be substantially offset by the improved marketplace power of users who are not locked in to a particular vendor's wares but instead obtain the benefits of continued competition for their business. The costs of technological diversity are real, but so are the benefits. This is true in non-disaster systems and even more so in disaster management systems where resilience is more important than efficiency. The diverse agencies responsible for responding to disasters and for acquiring technology to do so should embrace the natural tendency toward heterogeneity that this organizational reality engenders. The challenge is to work with other agencies to standardize the right things.

Another driver of purpose-built proprietary systems involves current efforts to eliminate technological stovepipes and single-agency systems that greatly hamper effective, interoperable communication on the scale necessary during a disaster. This is an important goal given the significant and ongoing problems with interoperability already discussed in Chapters 1 and 2. These systems improve interoperability but generally do so by moving the stovepipes to a higher organizational level (e.g., multiple police departments are integrated into a state-wide police communication system, or municipal police and firefighting communication systems are integrated within a municipality or region), but boundaries will remain (e.g., police and fire fighters cannot communicate, or public safety agencies across jurisdictions cannot communicate). Federated solutions made up of diverse technologies and platforms could be implemented without the technological and operational risks associated with purpose-built, integrated solutions—and still provide interoperability, as long as interoperability points are designed into the system.

Mobile phone companies have managed a major technology generation change about once every 10 years. Is this an argument for a monolithic system or a federated system? Mobile networks generally consist of monolithic and non-interoperable technology at the wireless interface. However, mobile phones have a clear interoperability point, the telephone network (and more recently the Internet). Operators may make technology changes (large and small) internally to their network while continuing to maintain an interface to the interoperability point. This is, in fact, why cellular networks built on Global System for Mobile Communication (GSM) technology can interoperate with Code Division Multiple Access (CDMA)-based networks. In the same way, different agencies might have different and incompatible networks but maintain an interoperability point with other agencies' networks.

Even within integrated, all-in-one systems, substantial value from technological diversity can be achieved by allowing for integration of diverse components built with standard interfaces rather than building end-to-end proprietary systems. Systems that have taken the trouble to meet the challenges of a diverse technology base at any one point in time are also "future-proofed" against technological change. New components become just additional instances of diversity, allowing incremental system upgrades that can leverage the latest technology advances, rather than introducing dilemmas requiring fundamental and potentially disruptive "forklift upgrades" or lost opportunities for improving effectiveness. That same flexibility makes diverse systems more agile in the face of sudden change. Mergers, relocations, and major changes in scale are characteristic of day-to-day operations as much as of disasters, and a diversity-tolerant approach to technology can facilitate them all.

## The User Interface and the Underlying Technologies to Deliver a Capability

Many familiar information and communication technologies are best known by the various controls, displays, sounds, or other sensory experiences they present to the user. The familiar package of a public safety responder's portable two-way radio stands, in many people's minds, for the whole infrastructure of frequencies, repeaters, and towers of which the radios are the outward and visible sign. First responder radios encapsulate multiple attributes—push-to-talk capability, the form factor of the handset and microphone, and communications within defined groups—that could be unbundled and repackaged for more effective use. Handheld communications devices might employ cell phone technology but with adaptations (in durability, form factor, or frequency).

But the association of the "user interface" with the underlying infrastructure is not nearly as fixed as in the past, and grows less so every year. Push-to-talk operation has, for example, been added to cell phone systems. Similarly, the Internet connects two-way radio networks and telephone calls, while telephones send e-mails and images to computers over the Internet. Television transmitters broadcast data, while TV programs are "streamed" over data connections. The long-heralded "digital convergence" of various communications and computing platforms has arrived, and in arriving has demolished many of the traditional connections between how a communications capability is embodied in a user interface and the underlying infrastructure used to deliver it. This is creating no small degree of confusion in procurement, regulation, and management of IT systems.

Too often, would-be innovators have been hamstrung by regulations or procedures that confuse the user interface with the underlying infrastructure. However, the opportunities created by tearing down the walls between systems are much greater than the costs. Single devices can be used to replace multiple devices, thus cutting costs and simplifying the work of first responders and emergency managers alike. Different systems can be interconnected in unforeseen and expedient ways, creating new capabilities such as circumventing damaged infrastructure.

The conflation of interface device (e.g., computer, phone) and infrastructure (e.g., the network hardware and networking protocols) also leads to the mistaken presumption that a diversity of communication systems must present huge challenges for the users to learn and effectively operate in a diverse technology environment. However, a variety of familiar as well as novel interface devices could coexist while still enabling interoperation and communication between systems—even where the underlying networks are vastly different. Likewise, users could be provided with a single, consistent interface that nonetheless utilizes more than one discovery, routing, and transport regime. The underlying infrastructure can hide the technological "glue" necessary to communicate between systems. Either way, interface device choices should be driven by how best to ease training and promote familiarity through routine use, and should be largely independent of decisions about the underlying infrastructure.

An awareness of the difference between these two distinct layers of technology and a thorough understanding of how they can be most effectively coupled with each other, depending on system requirements, is a critical skill for builders and managers of disaster management IT systems.

## THE TECHNOLOGY PIPELINE

The current state of the art in technology and level of technology adoption differs considerably across the capabilities discussed in Chapters 2 and 4. These differences in terms of readiness can be usefully thought of in terms of a technology pipeline. Where a given capability is in the technology pipeline affects the type of investment needed to advance it toward use in practice and the amount of time before a payoff can be expected. A technology investment process to produce a steady stream of improvements for incorporation of IT in disaster management should include a detailed assessment of where technologies are in the pipeline. Investments are most likely to provide continuous improvements when balanced across the spectrum of possibilities—from adoption, adaptation, and development to applied research and general research. A roadmapping process, described in the following section, is a tool that could help guide such investments.

One way of dividing up the stages in the technology pipeline is the following:

- *Adoption—technology available today requiring efforts to overcome adoption barriers.* Some IT technology is already available but has not been widely (though possibly partially) adopted by the public safety or disaster management communities that could benefit from it. Such technology does not require significant further adaptation, development, or research. There are many reasons that such technology has not been adopted, including cost, lack of awareness or training, and resistance to change. Nonetheless, adoption of these technologies can have immediate impact on attaining important goals for the use of IT for disaster management. The time to realize benefits from these technologies is limited chiefly by the time necessary to overcome the barriers to adoption, which will vary by organization and the nature of the barriers to be overcome. These barriers may include the time to secure the investment and to make the appropriate connection between that investment and the fully discounted net present value of the savings that the investment will accrue. Advances in general data processing, storage, communication, display, and software technologies will continue to make components of IT systems less expensive and more capable with time—aiding the adoption process. (Examples of technologies in this category and described in Chapter 4 include radio-frequency identification (RFID) for resource tracking and logistics; computer-mediated exercises; reverse 911 capability, i.e., two-way emergency reporting; and portable unmanned aerial vehicles and robots.)

- *Adaptation—technology on the horizon and ready for transfer to disaster management practice.* Effective systems to serve disaster management may be assembled by combining and adapting available commodity components. IT systems are mainly built using hardware that is available off the shelf and adapted using off-the-shelf software components. Some hardware components—such as personal computers and disk drives—have become commodities. Commodity hardware benefits from significant economies of scale, so that the hardware cost of a cell phone is roughly one-tenth that of a police handheld radio. Part of the cost differential is due to specialized functionality and ruggedness requirements, but a significant portion of the difference comes from much higher production volumes. Adapting commodity hardware to disaster management (in contrast to developing wholly novel hardware) could significantly lower cost due to specialized requirements. Software exhibits more flexibility. Many software and Web-based applications are expressly designed to allow customization for specific uses. Some useful software exists that has been developed in an open and freely shared environment that lends itself to adaptation and customization. In such an environment, investment can be distributed, often close to the end users, making it possible for users and vendors to adapt many existing IT technologies to disaster management readily and rapidly. Useful results can be obtained on time scales measured in months. (Examples of technologies in this category and described in Chapter 4 are commercial collaboration software and file sharing, online resource directories, multiple input/multiple output wireless systems, integrated ad hoc data collection tools (blogs/wikis), and mobile cellular infrastructure.)

- *Development—technology on the horizon and development needed for use in disaster management.* For some requirements the technology and design principles are fairly well understood, but existing technology is simply not adequate for disaster management. A concerted effort is required to develop significant software, hardware, and organizational structures to take advantage of them. In this case, a request-for-proposals process can be used to solicit capable organizations to deliver a product that implements the desired technology. Development time depends on project complexity, but useful results can often be obtained within a year. (Examples of technologies in this category and described in Chapter 4 are volunteer mobilization systems, event-replay tools, and intelligent adaptive planning tools.)

• *Applied research—issues requiring disaster-management-specific research.* There remain some difficult issues in disaster management for which solutions are not at hand—for example, reliable radio communications inside buildings or rubble. Research aimed specifically at improving disaster management could be conducted by university, commercial, and government laboratories, and even volunteer relief agencies such as the Red Cross. This type of activity is managed and directed within the government by agencies such as the National Science Foundation, the Department of Homeland Security (DHS), the National Institutes of Health, and by defense-related organizations such as the Defense Advanced Research Projects Agency and the service research laboratories. Support for smaller companies is given through Small Business Innovation Research and Small Business Technology Transfer programs administered by many of the above agencies. Because of the nature of disaster management and the types of challenges that the community faces, disaster-management-related IT research is becoming increasingly broad and interdisciplinary (see Box 3.3 for a discussion of the challenges of interdisciplinary research), involving contributions from multiple technical and social science fields. Fully realizing potential gains will often involve the fostering and management of collaborative research. Even so-called short-term research is typically a multiyear proposition and requires validation in the field at real disasters, as well as simulations. (Examples of technologies in this category and described in Chapter 4 are software-defined radios, tools for data mining across diverse information sources, decision sentinels, deployable sensor networks, and computer-assisted disaster simulation training tools.)

• *General research—issues requiring research followed by adaptation to disaster management.* Some problem areas in disaster management overlap general needs in IT management. IT is a broad, active area of research, and relevant research aimed specifically at disaster management is performed at university, commercial, and government laboratories and is sponsored by the same constellation of agencies. Many of these labs are engaged in broad areas of research that have the potential to develop new IT capabilities which, though not directed specifically to that end, could be harnessed for disaster management. As with applied research, this research is typically a multiyear proposition. Further development or adaptation may also be needed for effective utilization in disaster manage-

> **BOX 3.3**
> **Interdisciplinary Research for**
> **Enhancing Disaster Management**
>
> Interdisciplinary approaches to disaster management have been discussed for quite some time. As noted in the recent NRC report *Facing Hazards and Disasters*, interdisciplinary research (i.e., research that blends researchers, expertise, and tools from a variety of disciplines to address compelling and crosscutting problems) has been gaining prominence in almost every field of scientific endeavor, including disaster management research.[1] Indeed, the report cites earlier NRC work that describes four factors promoting the growth of interdisciplinary research: (1) the complexity of nature and society, (2) the desire to address scientific problems that cross disciplines, (3) the need to solve society's problems, and (4) the power of new technologies.[2]
>
> The benefits of interdisciplinary research for disaster management can be substantial. For example, *Facing Hazards and Disasters* describes a number of "exemplars" of interdisciplinary research in disaster management—from infrastructure failures and urban economics to casualty analysis through a common framework to decision making for risk protection. However, *Facing Hazards and Disasters* also goes on to describe how interdisciplinary research can be particularly challenging when overlapping social sciences with natural sciences (something one sees quite a bit of in disaster management research).
>
> According to the same report, interdisciplinary research for disaster management faces a number of significant challenges of its own (in addition to the normal challenges for such research, such as lack of funding and academic incentives) if it is to prove successful. For example, the report notes that some issues often stem from "the failure of a research team to function collaboratively" owing to such things as difficulties in spanning culture gaps between the disciplines or the devaluation or undervaluation by one discipline of the work of another. Another challenge cited revolves around how disaster management research is most often viewed as applied research rather than basic research geared toward advancing overall knowledge in a given area.
>
> *Facing Hazards and Disasters* surveys the available literature in the area and also suggests a number of factors contributing to the success of interdisciplinary research (pp. 186-187). First, it notes that problem-oriented research is probably best suited for interdisciplinary work in disaster management. Second, it notes that the particular characteristics and abilities of researchers—including such things as interpersonal skills—are very important for such interdisciplinary research. Third, it describes how studies that keep research teams relatively small and have stable membership appear to be more successful at integration and research.
>
> ---
>
> [1]National Research Council, *Facing Hazards and Disasters: Understanding Human Dimensions*, The National Academies Press, Washington, D.C., 2006, p. 181.
> [2]National Research Council, *Facilitating Interdisciplinary Research*, The National Academies Press, Washington, D.C., 2005.

ment, which may add more time. This report identifies topics that require general research, but it is not expected that funding specific for disaster management will be employed for these topics. (Examples of technologies in this category and described in Chapter 4 are delay-tolerant networking, automated information fusion from diverse sources, and calibrated information confidence tools.)

## ROADMAPPING

Any research agenda aimed at improving the long-term effectiveness of IT use in disaster management must be placed in the context of the technology pipeline and must prioritize the items in the agenda against each other, in particular noting where progress in one area may be dependent on progress in other areas or on organizational advances. An efficient approach to investment requires a clear vision of the path to improvement and a detailed understanding of the individual pieces of the problem and their interrelationships, together with a mechanism to measure progress.

Disaster management is, ultimately, a system-level problem. Thus, improving IT use in disaster management requires a system-level approach. The research agenda is likely to have the most impact if it conforms to a clear vision of the path to improvement defined in a fully articulated roadmap. By establishing a process for making improvements, currently unimagined concerns can be efficiently addressed as they arise, and as both technology and practice evolve.

A technology roadmap is a planning tool that can provide information to make better technology investment decisions by identifying critical technologies, technology gaps, and interdependencies between technologies that dictate coordination of research and development cycles. It can also help uncover interconnections between technologies and adoption issues related to organizational or human behavior characteristics. Perhaps most importantly, it can serve as a mechanism through which diverse participants, often with conflicting priorities but with common goals (i.e., saving lives and reducing economic and other impacts), can cooperate to address a larger problem of common interest—in this case the most effective handling of disasters possible.

Until relatively recently, the technology choices facing most disaster management organizations were comparatively few, with much of the investment focused on building specialized communications systems in close partnership with a small set of vendors. Today, there is a much wider set of technology options available. There is also an increasing need for the diverse organizations with public safety and homeland security responsibilities to be able to cooperate during large-scale disasters. In response, efforts have been made to identify appropriate technologies (such as DHS's Select Equipment List).

An effective, useful roadmap is driven by a clear set of user-driven (not technology-driven) goals and needs to evolve continuously as a living document in consultation with the full range of stakeholders. Some pieces of a roadmap are in place (e.g., the National Incident Management System), but an overarching strategic vision of how IT can best be evolved and applied to disaster management is missing.

A roadmap can serve as an anchor for a strategic vision and help policy makers avoid lurching from one priority to the next, driven by the most recent major disaster. Unfortunately, in the absence of a roadmap, more or less haphazard, reactionary IT investment is occurring and is likely to continue.[10] New incidents (like a major hurricane) should trigger a reevaluation of the existing roadmap, potentially leading to some adjustments in priorities within the strategic framework. But the continuity of investments should result in continuous and more predictable improvements in the application of IT to disaster management.

A roadmap can also assist policy makers and planners in balancing investments across different technologies appropriate at different times in the disaster management life cycle, that is, mitigation preparation, response, and recovery. A roadmap can also make explicit investment choices concerning tradeoffs among competing priorities and between tensions such as security versus openness and other such tensions previously identified.

Finally, a roadmapping process provides an opportunity to consider the interrelationships between technology and organizational models and technology and policy. Successful technology development and deployment are widely understood to require active consideration of the organizational context in which they will be introduced. Similarly, potential policy barriers must be considered when developing new technologies and organizational approaches.

Critical to the success of a roadmap activity is the inclusion of a broad array of stakeholders and an institutional home to get started and remain viable. All participants must make a long-term commitment to the resulting roadmap and to its continuing evolution as technological advances and organizational innovations are made.

---

[10]The fiscal year 2007 Senate Appropriations Bill for the Department of Homeland Security shows evidence of this type of planning, focusing on hurricanes and immigration. See Michael Arnone, "DHS Bill Slashes Research Funds," *Federal Computer Week*, July 17, 2006, p. 11; available at http://www.fcw.com/article95287-07-17-06-Print.

## Examples of Successful Technology Roadmapping

Roadmapping is a technique frequently used by firms to plan future research and development activities. The U.S. military, for example, uses a roadmapping approach in its Quadrennial Defense Review report[11] to drive plans for incorporating technology advances into its future capabilities. The International Electronics Manufacturing Initiative has developed a sensors technology roadmap that examines technology capabilities and applications in a variety of sectors, including transportation, health care, and consumer electronics.[12]

Perhaps the most familiar application of roadmapping is the semiconductor industry's roadmap. In the late 1980s, it became clear that the integrated circuit industry was not only a rapidly growing part of the global economy but also critically important to the economy and national security of the United States. Unfortunately, concerns grew that the United States, after an initial leading role, had fallen behind in technology leadership relative to other countries, especially those in Asia. A 1990 National Research Council report outlined the consequences of not maintaining a commercial and technological lead in this area.[13] The Semiconductor Industry Association (SIA) would take on the role of pulling together a long-term technology roadmap for the industry based both on end-user needs and technology trends.

This was not an industrial plan in the usual sense, but rather an agreed on, coordinated vision that would help each organization plan development and investment strategies that would bring the thousands of pieces of technology needed to make an integrated circuit together at the right time and the right level of development. A technology piece developed too early would be prohibitively expensive—developed too late it would not be profitable. Market forces would ensure that vendors and suppliers would tool up to meet a particular need at the right time, and even researchers understood the targets for conventional technology and could choose areas for research that, if successful, would have the most impact. Unlike other attempts at planning, this was not directed at a *specific* technology goal, but rather at the *process* of continuously improv-

---

[11]The 2006 *Quadrennial Defense Review Report* notes in its preface that the ideas and proposals in the report are provided as a roadmap for change. The report is available at http://www.comw.org/qdr/qdr2006.pdf.

[12]Charles E. Richardson et al., "Sensor Technology Roadmapping Efforts at iNEMI," *IEEE Transactions on Components and Packaging Technologies* 28(2): 372-375, June 2005.

[13]Computer Science and Telecommunications Board, National Research Council, *Keeping the U.S. Computer Industry Competitive: Defining the Agenda*, National Academy Press, Washington, D.C., 1990.

ing a key technology over a span of decades. Each step along the way would have important economic and strategic value and would form the foundation for the next important advance.

Built into the creation of the roadmap was a process that drove continuous updates and refinements, making it a living document with continued relevance and ensuring that it was up to date with advances in science, technology, and market needs. By the end of the 1990s, after three major updates and the associated strengthening of the U.S. semiconductor industry, there was a push to expand the roadmap process to a global scale so that the vision would match the expanding scale of the industry. Today it is a joint effort of industry, government, and academic representatives from the United States, Europe, Korea, Japan, and Taiwan. It is the critical common view driving ongoing investment in research, development, and manufacturing in one of the largest and most complex components of the global economy.

The roadmap for disaster management would be quite different from that of the semiconductor industry. Specifically, the SIA roadmap was made possible when an entire industry needed to plan for future generations of fabrication equipment and realized that this highly capital-intensive equipment was beyond the means of any one industry participant—it required cooperation with other participants. This created an environment where cooperation within a specific framework, embodied by the SIA roadmap, was possible, while allowing continued competition in areas outside of that framework.

In contrast, disaster management organizations and the associations that represent them would necessarily drive the roadmapping process envisioned here. Yet, the key element remains—the need to create a framework within which cooperation can happen in order to address common goals that are otherwise unattainable or suboptimal. By joining together to develop a roadmap, they would have a forum for speaking with a common and consistent voice to the vendor community about technology needs.

There are also similarities from a process perspective; stakeholders create a living document that explicitly lays out a vision for continuous progress based on balancing value and cost, as well as carefully considering technical and organizational feasibility. Then, investment from all sectors can then be committed to track this vision as it evolves. A successful roadmapping process would ultimately result in full and active participation of the vendor community, just as the SIA roadmap process eventually included the entire worldwide semiconductor industry, including those who spurred its development as the perceived "adversaries."

## RESEARCH CENTERS: COUPLING TECHNOLOGY RESEARCH WITH PRACTICE

Successful development, adoption, and utilization of IT for disaster management require several different communities be in regular and close contact with one another. Researchers tend to look for overarching themes, but experience has demonstrated the importance, in the field of disaster management, of starting with real problems faced by real practitioners, working back from there to overarching themes. Starting with overarching research themes will likely lead to dead ends and unimplementable technology.[14] Practitioners must help define needs for new technology, thus acting as inspiration for researchers and developers. They must interact with developers and vendors throughout the prototyping cycle and development process to ensure that their needs are indeed addressed. IT researchers must have opportunities to expose practitioners to novel concepts in order to generate an understanding of potential new capabilities and how they might fit into current and future operations. Public administrators, social scientists, and IT researchers all play important roles in ensuring that IT innovations are introduced with the necessary organizational changes to enable new devices and systems to be smoothly integrated into practice.

Forging organizational ties is harder in disaster management than it is in sectors like defense because the vast majority of practitioners are distributed across local agencies that are normally fairly isolated from each other and from the research community. Nevertheless, integrating the experiences and needs of these different agencies is crucial, since sooner or later when a disaster of a severe enough magnitude strikes, they are bound to have to work together. Some regional groups of organizations that have already experienced the need to work together have successfully initiated the process of forging organizational ties, suggesting that building from a bottom-up approach is likely to be most effective.

Moreover, successful IT development is iterative. It is important to provide practitioners with initial prototypes to bootstrap the iterative process. Testbeds and exercises are particularly critical in the area of disaster management because they provide opportunities for feedback from actual users about critical requirements of responders that may not otherwise be apparent. In some cases, large-scale testbeds are required to understand issues that only emerge at scale. Simulations present oppor-

---

[14]This point was made repeatedly by workshop participants and is reflected in the committee's earlier workshop report; see National Research Council, *Summary of a Workshop on Using Information Technology to Enhance Disaster Management*, The National Academies Press, Washington, D.C., 2005, p. 7.

tunities not only for training but also for observation and assessment of IT capabilities such as decision support tools. Operational facilities that permit instrumentation, experimentation, and iteration are needed.

Collaborative research centers could, therefore, play a highly useful role in advancing the effective application of IT to disaster management. The major goals of such centers would be sixfold—(1) to develop a shared understanding of the experiences and challenges in all phases of disaster management from both a technological and organizational perspective, (2) to evaluate the application of technology advances to disaster management practice, (3) to develop a culture and processes for transitioning knowledge and technology to the operational communities on a sustained basis, (4) to build human capital at the intersection of information technology and disaster management, (5) to serve as repositories for data and for lessons learned from past disasters and disaster management efforts, and (6) to provide forward-looking analysis to inform the development of technology capabilities, associated organizational processes, and roadmap development.

The research conducted by these centers would be multidisciplinary, combining the efforts of information scientists, engineers, and social scientists. Participants would be charged with collecting knowledge and experience from past disasters and using it to build a core set of knowledge that would inform the development of technology capabilities and associated organizational processes to enhance the management of future events. The centers would closely partner with federal, state, and local agencies responsible for disaster management. Indeed, experienced and capable emergency management officials and operational units from disaster management organizations should be deeply involved in the work of these centers. One approach for engaging these government agencies could be to provide them with incremental funds specifically for working with researchers and to develop next-generation technologies.

To ensure that the work of the centers is informed by and responsive to the needs of disaster management, centers would bring in disaster management professionals from all levels of government as visiting fellows. To inform additional researchers about the problems of disaster management, university faculty and students would be offered internships and fellowships. Finally, to help encourage development of technology based on the research results, the involvement of relevant industry would be promoted through informational activities and the sharing of expertise and results.

Multiple centers for research would have several advantages over a single research center. They would enable healthy intellectual competition and cross-fertilization of ideas and allow for specialization in specific types of disasters, specific technology capabilities, or the comprehensive needs of particular geographical areas. Certain research centers could, for example, specialize in disasters common to their locations, in order to benefit from expertise residing in local emergency response organizations and other local government agencies. For instance, a center near known earthquake-prone areas may focus on technology related to improving earthquake-specific disaster management. Different centers could specialize in practical and response-oriented work, combining core as well as geography-specific expertise. Close coordination and sharing of information and expertise among centers would help avoid unnecessary duplication.

A major goal of these centers would be to develop a culture of continuously transitioning knowledge and IT between researchers and operational communities. This is very different from the usual academic model of licensing technology to a third party, or creating a start-up. A continuous process of reviewing user requirements, knowledge generation, collaboration, validation, acceptance, implementation, and incorporation of new user needs must be encouraged.

Field research—working on large problems outside the labs—appears to be particularly valuable to making progress on using IT for disaster management. It pushes researchers in new directions. It also stresses the technology under the extreme conditions inherent in disaster situations, exposing issues unlikely to be discovered in a laboratory setting. Practitioners' participation in such research gives them an opportunity to see the potential for new information technologies. It also gives them a chance to influence its direction. The goal is to close the gap between researchers and practitioners and create a unified core community that can speed up the process of delivering research results of immediate relevance to disaster management. Panelists at the workshop held by the committee cited the Disaster Management Interoperability Services (DMIS) Program and the Biological Warning and Incidents Characterization (BWIC) projects as successful examples of programs carrying out field research that involved the public safety community.[15] The Strong Angel exercises mentioned in Chapter 2 are another example of how technologies still in the development stage can be tested in the field and can begin to gain acceptance in the practitioner community that is ultimately indispensable to adoption, as well as provide researchers with feedback on the proper direction for further research and development.

Finally, as the use of advanced sensors, communication technology, and similar IT increases, it becomes ever easier to collect data about the

---

[15]See National Research Council, *Summary of a Workshop on Using Information Technology to Enhance Disaster Management*, The National Academies Press, Washington, D.C., 2005, p. 6.

process of dealing with a disaster in a completely unobtrusive manner. Such data ought to form a basis for studies that will ultimately lead to improving the disaster management process, and it should be used to help evaluate new proposed technologies and methodologies. Centers should serve as repositories for these data.

Several research centers devoted to certain aspects of disaster management already exist. Some well-known centers are the Natural Hazards Center at the University of Colorado at Boulder; the Disaster Research Center at the University of Delaware, which investigates the social science aspects of disasters; the Hazard Reduction and Recovery Center at Texas A&M University; and Dartmouth College's Institute for Security Technology Studies.[16]

Such centers could provide the basis for a network of research centers where IT researchers, hazard and disaster researchers, and disaster management practitioners can collaborate to study and evaluate the use of IT for disaster management from both a technological and an organizational perspective; transition knowledge and technology to those who practice disaster management; build human capital at the intersection of IT and disaster management; and develop future IT capabilities.

---

[16]Texas A&M University provides a Web site at http://archone.tamu.edu/hrrc/related-sites/Centers.html#Domestic with links to domestic and international disaster research centers. The Natural Hazards Center at the University of Colorado at Boulder provides links to Web sites of U.S. and international organizations dealing with hazards and disasters, including academic research centers; see http://www.colorado.edu/hazards/resources/centers.

# 4

# Elements of a Research Agenda

The six key information technology (IT)-enabled capabilities discussed in Chapter 2 represent broad areas where there is significant potential for the application of IT to enhance disaster management. These IT-enabled capabilities are used here to provide a framework for organizing research and development needs and opportunities described in this section as part of a research and development agenda. This chapter is intended to provide an initial sketch of the kinds of items that would ultimately appear in an IT roadmap for disaster management. The committee sought in particular to identify technologies under investigation that may hold largely unrecognized promise for advancing disaster management, though more apparently promising technologies are described as well. Developers of a roadmap could use this survey as a starting point for developing a fully articulated plan, including a detailed set of research directions to be pursued.

In developing this initial sketch of a research agenda for use in an IT roadmap, the committee made some assumptions about the continuation of a number of technology trends occurring independently of the needs of disaster management. The agenda is not aimed at influencing the direction of these trends, which the committee believes will continue regardless of the research agenda identified here. However, an IT roadmap would have to assume the continuation of current trends as a necessary foundation for the development and commercialization of major aspects of the disaster management research agenda and should incorporate them as base technology trends. Roadmap developers may need to account for

ELEMENTS OF A RESEARCH AGENDA 109

deviations from those trends should they occur. The technology trends assumed are as follows:

- Continued improvements in the cost and capabilities of all aspects of computing and data processing, including computer processing power ("popular" form of Moore's law), storage capacity, energy efficiency, and network bandwidth;
- Continuing increases in the capability in the existing commercial cellular network: more capacity for voice and data, as well as new classes of handsets with long-battery-lifetime data capability, and increasing functionality;
- Continued progress toward transforming communication systems from vulnerable and unreliable direct links to more robust, distributed, and self-repairing grids using new approaches in mesh networking and adaptive radios;
- Continued progress toward universal compatibility of *all* communication and computing technologies by basing them on the underlying Internet technology (i.e., "IP everywhere");
- Continued growth of open source software methodology to allow rapid creation of new, high-reliability applications built on the experience base of existing ones;
- Increasing use of high-volume commercial technology components in specialized systems in order to spread component development costs over a much larger base, to build systems more rapidly by exploiting existing components, and to increase reliability through increased field experience;
- Rapid spread in the availability of low-cost, ubiquitous public wireless data networking, especially at the municipal level;
- Continued improvements in linking geographic information to existing and future data applications (e.g., Google maps), and all types of databases such as geographic, medical, building plan, toxicology, weather, and image databases; and
- Continued rapid progress in biochemistry, electronics, micromechanics, and nanotechnology that can be expected to lead to cheaper, smaller, and better-performing sensors, actuators, and other devices.

Under each key IT-enabled capability, technologies are listed roughly in order of their current position in the technology pipeline. This is necessarily only an approximation, as many technologies consist of multiple iterations and incarnations that vary in their degree of readiness (and value) for application in the field. For each item listed, a brief description is provided, indicating both general and specific directions for potential advances and some example applications for disaster management.

Some of the technologies listed with one capability area may apply across a number of capability areas. They have been listed with the capability area that they appear most likely to advance. Similarly, many of the technologies are interconnected. They may need to be combined to effect improvement in a specific application, such as better assessment of structural damage. Advancement in one area may be predicated on advancement in other areas: for example, improvement in energy technology to power mobile devices is required before advances in several other technology areas can be widely exploited. This report does not make a detailed assessment of these interdependencies. However, identifying and planning for them would be a major part of the development of a technology roadmap.

## MORE ROBUST, INTEROPERABLE, AND PRIORITY-SENSITIVE COMMUNICATIONS

- *Exploitation of cellular, wireless networking, and Internet Protocol (IP)-based technology.* All of these commercially available technologies can be applied more systematically in disaster management practice to improve communications resilience by adding redundancy and standards-based interoperability. Effective use of these technologies requires enhancing operational procedures among public safety and emergency management organizations. They should seek to employ alternative communications capabilities and implement fallback communication strategies when communication resources are overloaded or impaired (e.g., text messaging in place of voice calls). The opportunities for integrating mature and maturing technologies into disaster management are abundant. The struggle is not so much the usefulness of the technologies, but the processes and structures, which limit their implementation. Box 4.1 details characteristics of cellular technology that hold potential for improving disaster management.
- *Redundant and resilient infrastructure.* Many communications problems are caused by the destruction of communication devices or communication lines and by the loss of power. These are often the result of damage to physical structures (buildings, cell towers). Hardening the infrastructure is an obvious way to reduce failure of equipment and lines and reduce the chances of power loss. While it is not economically infeasible to harden all relevant equipment for worst-case scenarios (worst-case hurricanes and earthquakes, terrorist attacks), improvements are certainly possible.
- *Mobile cellular infrastructure.* Cellular system capabilities could be modified and satellite communications links could be integrated to enable quickly deployable communication systems for use in disasters.

- *Intelligent spectrum sharing.* Spectrum shared more flexibly between commercial and public safety users and better use of spectrum through improved radio technologies and policy continue to receive significant focus. It is for this reason that this report mentions spectrum issues only in passing, though the committee views dealing with spectrum issues as an important component of any technology roadmap.
- *Multiple input/multiple output (MIMO) wireless systems.* Antennas, coding, and modulation could be improved to enhance radio performance. MIMO wireless systems take advantage of miniaturization to incorporate an array of radios and antennas in both the handsets and base stations as an economical way to increase user capacity. This technology has demonstrated a multiplicative increase in capacity and spectral efficiency; dramatic reductions of signal fading thanks to diversity; increased system capacity (number of users); and improved resistance to interference. Fundamental research is needed to fully realize these benefits in practical wireless systems.
- *Non-voice communications for first responders.* Voice is a natural communication medium, and first responders are trained to use it reliably under stress. Unfortunately, even the best-trained person is limited in the amount and kind of information he or she can process using voice communication. Careful integration of voice input/output with data applications could extend users' ability to process information from a broader knowledge base while reducing cognitive overload to leverage the scarce asset of trained operators.
- *Integrated voice/data/video.* Separate systems for different modes of communication are inherently inefficient (although separate systems may provide useful redundancy), and integrated communication provides economies of scope and scale. By providing users with a more general purpose device, new applications can be more easily introduced. Many technologies (e.g., IP-based, cellular) naturally provide both voice and data communications in an integrated fashion: that is, they use a single infrastructure, set of protocols, and terminal device. Adaptation of the technology, such as the development of special terminals that meet the robustness requirements of first responders or that also provide access to existing communication infrastructures, may be required.
- *Policy-based access control mechanisms.* Access control lists (ACLs) are used extensively today for enforcing policies governing the type of access (e.g., read, read/write, print) a user has to information. Policies are hard-coded into ACLs, but having the policy hard-coded causes management and security problems. Research into mechanisms for implementing ACLs that eliminate these problems promises to provide the flexibility and ability to operate in the inherently dynamic environment of a disaster.

> **BOX 4.1**
> **Opportunities to Better Apply a Maturing Technology: Commercial Cellular Technology**
>
> Cell phone technology has several salient features today for communications in a disaster. Cellular technology provides communication that is logically defined. Whom one communicates with is defined by who is dialed or by logical talk groups rather than by who happens to be within radio range. This implies that one can exclude conversations within radio range as well as communicate with users beyond this range via networking technology.
>
> Cellular technology communicates with other technologies, both wireless and wire-line. Cellular communication interconnects with many different cellular radio technologies, with traditional landline phones, and with Internet-based phones. Both point-to-point communications and logical talk group communication is possible. Nextel using Motorola iDen technology has led the way with walkie-talkie-like talk group technology. Other cellular technology vendors are following suit. Cell phones send both voice and data. The data include short messaging service, Internet access, and communication control data used to operate the system.
>
> Applications other than voice are already available. E-911 provides a localization application. (Many E-911 implementations use the Global Positioning System [GPS]; other technologies are available.) Camera phones can take pictures or short videos and send them. Cell systems can have information services that broadcast local weather, traffic conditions, breaking news, and so on.
>
> The hardware cost of cellular handsets is low relative to that for land mobile radio (LMR) handsets, in part owing to the economies of scale from so widely adopted a technology. Cellular handset battery life is quite good. A phone with low activity can last many days. A busy phone can have hours of talk time.
>
> Cell phones have many inherent personal digital assistant capabilities, such as memo pads, alarms, and searchable phonebooks. Cell phones are small and lightweight, a fraction of the equivalent LMR radio. The keyboards are small, but many handsets include hands-free voice dialing. The handsets are designed to be quickly reprogrammed with different cell phone numbers, service provider profiles, and network configurations. In some cases this can be done over the air. Cell phone interfaces come in packages other than handsets and can be embedded in

- *Software-defined radios.* Software-defined radio (SDR) technologies provide software-controlled configurability of radios and their incorporation into networks. Remote configuration of frequency bands, interference management, operational and control protocols, and security options offer the promise of flexibility, easy incorporation across heterogeneous organizations and applications, and life-cycle cost reductions through high volume. (See Box 4.2 for further discussion.)
- *Delay-tolerant networking.* Current communication systems that depend on point-to-point links are vulnerable to disruptions but have the advantage of very low delay when they are working. Distributed systems are inherently more robust because they can reroute traffic as needed, but

> equipment and other devices. For instance, a sensor might have use a cellular interface to report data, or a tracking tag could use the cellular infrastructure.
>
> Cellular can operate robustly under high traffic loads. When the load of the system exceeds capacity, calls will have to be blocked. The question is whether the network still maintains control over who gets blocked and what kind of communication can take place. This depends on the communication control mechanisms. Most cellular standards automatically expand their communication control capacity as the demand on the network increases. They can maintain control of the network with network overloads that are many times the capacity of the traffic channels. To a common user, voice calls will be blocked and the system unusable. Other users can be given priority access to the available channel and may experience little or no blocking. Though voice calls are blocked, the common user may still be able to send other types of communication such as short message service, a phenomenon noted after the July 2005 London bombings and after Hurricane Katrina. These features of cellular technology can be exploited for disaster management. The difficulty is whether this is done using commercial cellular infrastructure or building a parallel or supplemental infrastructure.
>
> Cellular technology is, of course, exploited for commercial purposes by cellular service providers. Commercial cellular service is provided throughout most populated areas. The cellular industry expends billions of dollars per year to expand the number of cell sites.[1] The area of coverage and capacity continues to grow steadily, along with the growth of subscribers and the types of services to which they subscribe. Building parallel infrastructure for emergency services is clearly cost-prohibitive. Yet, both public safety communications officials and cellular infrastructure owners have valid concerns about piggybacking disaster communications on existing commercial infrastructure. Addressing these concerns and overcoming the issues with innovative policy approaches in the spirit of public/private partnership should be a major focus of roadmap developers.
>
> ---
>
> [1]According to the Cellular Telecommunications Industry Association, a total infrastructure investment of $25.3 billion was made in 2005 alone. Other statistics on coverage and capacity are available at http://www.ctia.org/research_statistics/index.cfm/AID/10202.

this leads unavoidably to longer and less predictable delays in transit. Delay-tolerant network (DTN) architectures and protocols are an effort to address this problem in IP-based networks (e.g., the Internet), particularly those operating in environments characterized by very long delay paths and frequent network partitions, such as mobile or extreme environments that lack continuous network connectivity typical in disasters. A focus of DTN research is to provide interoperable communications in such environments.

- *Passive and active embedded links and relays.* These devices hold promise for enhancing communication in buildings, rubble, and underground. Building construction has been modified and existing buildings retrofit-

> **BOX 4.2**
> **Software-Defined Radios**
>
> Advances in digital technology are making possible a new generation of communication devices that can be configured in software to operate on different frequency bands, in different modes, and to present different operational and control options to the user. Such "software radios" can be reconfigured remotely ("over the air"), enabling the rapid reassignment of wireless devices among different pre-existing networks. Rapid reassignments can provide several important end-user capabilities. First, allowing the radio to switch between many standard modes will facilitate connectivity between incompatible networks and allow individuals to move smoothly between task groups operating on different legacy systems.
>
> Second, adding sensing and reasoning capabilities creates so-called cognitive radios that can dynamically adapt to the current user traffic needs, radio congestion, spectrum and operational policies, and radio capabilities to optimize overall communication efficacy; the addition of these capabilities is a precondition for several novel proposals for expanding the spectrum available to first responders in disasters. Finally, in extreme conditions, such as high noise or poor propagation through a building, the user can select radio modes and frequencies that will optimize their connectivity. Because of the critical importance of power and cost to disaster management communications systems, there should be a limited number of well-regulated modes for such radios to use, as determined by both research and current practices. This would allow mobile units to be optimized for efficient use of each of these modes.
>
> The more attractive option of having a completely flexible radio that can process an arbitrary signal will probably remain impractical because of the energy requirements of complex computations at gigahertz frequencies. Many challenges remain before such systems will be available, such as how various radios get informed as to which traffic they should copy and which traffic they should ignore, and how that gets done dynamically, in real time, and securely (if necessary).

ted with technology improvements for better earthquake survivability. Similar advances are possible to improve the resilience of communications infrastructure. These advances may allow wireless communications into and out of the building and through the rubble of a destroyed building, through passive means such as embedded tuned conductors or active means such as embedded relay arrays, similar to the emergency lighting systems already mandated.

- *Policy-based routing and congestion management.* Transmission congestion occurs during periods of high volumes of communication when information arrives faster than the network can handle it. Policy-based routing and congestion management could make it possible to prioritize and schedule transmission according to a specified policy. A policy-based framework would permit the behavior of the system to be modified without having to re-implement it and would also allow automated response to deal with congestion.

- *Self-managing and repairing (autonomous and adaptive) networks.* Autonomous and adaptive networks that are able to reconfigure and repair themselves could improve infrastructure robustness, even when infrastructure has been partially destroyed. These networks are envisioned to be unattended systems, with applications relevant to disaster management ranging from environmental sensing to structural monitoring and emergency response. Advances may result in communication infrastructure that supports need-based connectivity without boundaries and with minimal human intervention. Such a capability requires research not only on low-level communication (i.e., connectivity) but also on connectivity management and adaptive computer-based learning and management systems.

- *Location-based monitoring of communications infrastructure capacity and adaptive access to available communications channels.* IT systems with these capabilities would allow matching of the message form to the fastest, most appropriate network environment available to each user. Automated routing could allow senders to put out a message and let the system figure out how to get it to the right recipients.

- *Speech interfaces.* Visual displays and input schemes that are familiar in the office environment are not always suited to the needs of disaster responders. In particular, the use of visual displays in the field, particularly in highly mobile and/or high-stress activities with difficult visibility conditions, can create unwelcome distraction and even induce a form of disorientation that can be extremely risky. As speech-recognition and speech-synthesis capabilities improve, speech-command and audio-output interfaces become increasingly feasible.

- *Tactile interfaces.* These are user interfaces that employ touch for input and/or output. Examples of existing tactile interface technology include a Braille reader and touch screens. In hostile and loud environments, auditory contact may be difficult. Advances in tactile interfaces, like the vibrating ring on a cell phone, offer promise for improving communications in these environments.

- *Adaptive mesh networks.* Although data networks are frequently described as self-healing in case of damage or disruption, many actual networks have been constructed more for economy than for resilience. Compromises have led to the creation of single points of failure and relatively inflexible routing schemes. Wireless "ad hoc" meshes would include the capability of rapid provisioning of voice and data service in disaster-stricken areas, a feature that could also enhance the reliability and capacity of day-to-day networks. If wireless ad hoc mesh networks were coupled with advanced, distributed machine-learning algorithms, these networks could continuously optimize their capabilities with minimal human intervention, minimizing the cost and delays associated with that. Adaptive mesh networks also offer the opportunity to use distributed computing techniques for increasing analysis and decision-making resources beyond those available to any single node.

- *Standards as "middleware."* Various terms such as "service-oriented architecture" and "enterprise service bus" are used to describe a data-interoperability strategy based not on centralized interconnection services, but rather on direct connections between diverse IT systems using standardized data representations. This is basically how the Internet works. Such intermediary data standards can enable interoperability among data systems without reengineering the systems themselves or adding new layers of technology. (See Box 4.3 for a more detailed discussion of service-oriented architectures.)
- *Advanced power sources.* Improvements in power sources including fuel cells and tritium could have profound effects on the ability to deploy and use IT in disaster management. Often, the lack of infrastructure-based power, the limited life of batteries, and the bulkiness of "portable" batteries or generators limit the scope and range for application of IT. Any chemical fuel generator or storage system—battery, fuel cell, turbine, compressed gas, flywheel, and so on—has basically the same fundamental energy limit of the chemical binding energy between atoms. Battery capacity is increasing slowly as these devices are better engineered, but major advances in chemical batteries are highly unlikely. (Box 4.4 describes these technologies in more detail.)
- *Mobile power source efficiency and energy management.* At least five techniques could be employed and advanced to improve the efficiency of powering mobile devices: harvesting energy from the environment, using passive radios that modulate reflected radio signals rather than generating radio signals directly, reducing interference by better traffic coordination, improving power source recovery and redundancy, and efficient spectrum sharing. (Box 4.5 elaborates on each of these techniques.)

## IMPROVED SITUATIONAL AWARENESS AND A COMMON OPERATING PICTURE

- *Radio-frequency identification* (RFID) technology holds promise in a number of areas critical to disaster management. Early prototypes exist of systems for tagging victims with treatment and other information useful to medical responders. Longer-range RFID tags and readers will make it possible to continuously track victims as they move through the system from evacuation to treatment centers. Tagging of assets will also aid in

> **BOX 4.3**
> **Service-Oriented Architectures**
>
> Service-oriented architectures (SOAs) model computing applications as networks of "black box" functions, each characterized by a published network interface based on shared technical standards. Instead of the tightly coupled integration of software modules on a single computing platform, SOAs build applications by selecting from among a range of services over a network in order to support a needed function or business process. SOAs promise numerous benefits, including several of particular interest to disaster management:
>
> - The ability to utilize multiple sources of a single service, which can help eliminate single points of failure due to disaster effects;
> - The ability to scale up applications efficiently during sudden surges in demand by concentrating resources on the particular services that present bottlenecks;
> - A reduction in the "switching costs" associated with migration from one provider's services to a competitor's, thus maintaining competition, reducing "lock-in" effects, and "future-proofing" the application as a whole; and
> - Enabling support of exceptional or ad hoc business processes by the rapid assembly of services into new configurations (sometimes called real-time integration).
>
> However, as SOA-based applications evolve to mirror business practices more closely, the actual design and evaluation of such applications become less and less a technical problem and increasingly a direct concern for the end users. Consumers of SOAs are increasingly called on to perform rigorous analyses and documentation of their requirements and practices. This can create conflicts with resource-bound emergency managers, who may perceive such rigor as a threat to their autonomy and their freedom to adapt to circumstances. A great deal of research remains to be done on the interface between service-oriented IT architectures and their users. Of particular importance is determining the most convenient and rapid methods of collaborating with end users in the processes of SOA design, evaluation, and adjustment.

resource discovery and tracking, allowing real-time views of deployment and status that can be used for optimizing effectiveness.

- *Embedded, networked sensors.* As the density of a sensor network increases due to sensors' reduced size, power, cost, and ubiquitous connectivity options, the data they return change in quality as well as quantity. Complex or subtle spatial and temporal patterns become visible that could not be resolved before. At the same time, increased data processing power and enhanced presentation techniques allow the analysis and fusion of these dense data streams into usable situational awareness and analytic products.
- *Routine information fusion.* Often, combining (fusing) information

## BOX 4.4
## Advanced Power Sources

The foundation of a reliable wireless communication system, whether built using mobile units or remote sensors, is a reliable and long-lasting power source. The complex networks envisioned in this report will require a means to provide that power in a lightweight, long-lasting package that can be recharged, if necessary, onsite. In addition, to the extent that it is desirable to move from simple voice communications to transmission of more complex data at higher rates, the energy requirements will increase even further. Over the past several decades, the aggressive shrinking of electronics has reduced the cost, size, and power of electronic devices at a rate of nearly 50 percent per year.

Unfortunately, over the same time interval, chemical batteries have been improving only slowly, with average specific capacity increases of less than 7 percent per year. Since there are no new elements available, the ultimate performance of chemical batteries is well understood, and the slow year-to-year improvements come only from steady engineering advances. Many alternatives have been proposed, from fuel cells to flywheels, but ultimately they all depend on the strength of chemical bonds and will not be able to provide truly dramatic improvements in capacity.

To improve the capability and reliability of portable IT equipment dramatically, there appear to be three possibilities beyond the small improvements expected from steady engineering of existing batteries. The first is to make full use of the rapid advances in the performance of electronics to reduce the power required for each function. This is critically important for computation functions such as coding or data acquisition, but less important for the communication function where the energy is consumed by generating a strong-enough radio signal to reach the intended receiver. The second route to improvement is to design the communication network from the ground up to be energy efficient. For example, a higher density of readily deployed small base stations or repeaters will allow the handsets to reach the network with a minimum of radio power. Also, changes in protocols, coding, and signal processing can make substantial improvements.

Finally, abandoning chemical power sources and adopting some form of nuclear power can provide substantial improvements, since the energy per atom available in a nuclear process can range from several thousand to several million times that for chemical energy. The first generation of such cells has energy content greater than 10 times that of the best chemical cells, and the theoretical limit for the useful life of these cells is at least 100 times that of chemical cells. This technology is based on the use of tritium, a radioactive isotope of hydrogen that can be easily encapsulated, is not chemically toxic, and emits only beta radiation that cannot penetrate the skin. This technology has been in commercial use for years to power emergency lighting and provide other specialty-task illumination needs where an external power supply would be unacceptable. Moving to other radioactive materials, like those already incorporated in smoke detectors, can increase the limiting energy densities by another factor of 1,000 or more, though the engineering problems of gathering that energy efficiently are not yet solved.

> **BOX 4.5**
> **Mobile Power Source Efficiency and Energy Management**
>
> Following are five techniques for improving the efficiency of energy management in mobile devices:
>
> - *Harvesting energy from the environment.* Energy harvesting has an important place for distributed sensor or repeater networks without external power sources where energy can be collected from low-power sources such as light, vibration, pressure, and so on and stored over long periods for application during emergencies.
> - *Passive radios.* Passive radios are currently used for low-bit-rate, short-range sensor systems, like inventory control tags, where the base station sends a signal to the sensor that just modulates the reflection of that signal back to the base station. The entire operation of the sensor can be powered by the signal from the base station, or a battery can be used to supplement this energy, but the battery has a very long service life because it does not need to transmit, only modulate, radio frequency.
> - *Reducing interference and noise.* At the system level, there is enormous potential to improve radio power-consumption efficiency through a properly designed communication system. Energy to transmit data depends on the ratio of signal power to noise (or interference) power. Lowering the noise or interference by an order of magnitude reduces the signal power requirements by an equal amount. (It is easier to communicate in a room full of people if each person speaks in turn rather than all at once.) For example, a typical public service radio (e.g., XTS5000 Motorola) has a range comparable to that of a cell phone (e.g., Motorola V276) but a performance that is substantially different. The cell phone can be used to talk/listen for nearly 4 hours or wait for an incoming call for nearly 300 hours, all on a battery weighing 1 1/2 oz. The public safety radio has a useful battery life of only 8 hours (about 30 minutes for talking, 30 minutes for listening, and 7 hours for waiting for a call) using a battery over 4 times heavier. This is partially due to a more efficient radio and partially due to the ability of the base station to tell the handset to lower its energy consumption dynamically to the smallest value sufficient for efficient communication.
> - *Efficient spectrum sharing.* Proper coordination and the use of narrowband channels could also lead to significant improvements in power requirements. The public safety radio blocks other radios from using its channel, in contrast to the cell phone system, which is designed for highly efficient sharing of scarce spectrum resources.

from several different sources can provide greatly enhanced understanding of a situation. An example might be a map showing both road conditions and the number of people needing evacuation across a geographic area. Unfortunately, such information sources were usually developed independently of each other for different uses in non-emergency conditions, and bringing them together is a significant, ongoing challenge in development and applied research, depending on suitability of source.

- *Publish/subscribe information dissemination.* One of the great challenges in a fluid situation like disaster management is to provide the right information (and the right amount of information) to the right person at the right time. Publish/subscribe network services enable information dissemination across a potentially unlimited number of geographically scattered publishers (information providers) and subscribers (information receivers) along with the ability to customize information at the level of the individual user. More sophisticated approaches would help to alert users to novel information identified as potentially relevant on the basis of those users' demonstrated patterns of interests.
- *Semantic routing.* Traditional methods of information routing are based on explicit designation of the recipient or recipients. In highly dynamic networks, this can create a "discovery bottleneck" as new participants join and existing participants shift roles. A frequent result is that key actors inadvertently are left out of the loop for essential information and others are overloaded with irrelevant data. The next step beyond publish/subscribe is to use semantic routing techniques that allow information to be shared on the basis of the users' content requirements as well as the originators' directives. Semantic routing coupled with intelligent workflow could offer the ability to move information proactively in order to accelerate processes (e.g., obtaining of permissions).
- *Data quality and availability.* Variations in data quality and availability often plague attempts to share and integrate information for decision making. Improving the completeness, timeliness, accuracy, and consistency of data are all areas requiring advances. At the same time, it is necessary to recognize that imperfect data are and always will be present. There is thus an abiding need for methods of measuring, expressing, and correctly interpreting "fuzzy" and intermittent and uncertain data.
- *Data mining across diverse information sources.* The challenge in an information-rich environment is the ability to extract meaningful knowledge from a plethora of information sources. Data fusion across diverse information sources has already shown its effectiveness, particularly in geographically based applications. The next step is to create analysis systems that can extract data and automatically look for patterns using information held in diverse, incompatible databases. An example would be to search for patterns in hospital admissions based on information including weather patterns and possible sources of toxin release. Data mining can find patterns obscured by the volume or diversity of data. Intelligent approaches can corroborate tenuous patterns by consulting multiple alternative sources guided by the nature and context of the conclusion.
- *User-centered situational awareness information presentation.* With advances in information collection, fusion, dissemination, and delivery, it may become possible to create customized and dynamic presentations for

each user with information most relevant to him or her in the most useful format given the user's location, responsibility, and changes in the situation. Research supported by the Defense Advanced Research Projects Agency's Improving Warfighter Information Intake Under Stress program offers some examples of the type of advances that may be possible.[1]

- *Optimized presentation of data.* IT systems could be engineered to adapt to such things as message form or content, the receiver's device form factor, or the real-time situation of the recipient and to optimize the presentation of data accordingly. Such capability could allow adaptive access to available communications channels, allowing matching of the message form to the fastest, most appropriate network environment available to each user. For example, the system might not send a complete geographic information system (GIS) file to a personal digital assistant, but it *would* send the complete file to a more powerful workstation with a high-resolution display; it could send a voice announcement instead of a text message to a fully garbed firefighter engaged in a rescue; and automated routing could allow users to permit the system to figure out how to get messages to the right recipients formatted for the equipment available to them.

- *Deployable sensor networks.* It is impractical to put all classes of sensors in all possible locations before a disaster. However, existing sensor networks (e.g., sensors in commercial cell phones, public transportation, or private vehicles) could be integrated with ad hoc sensors with appropriate capabilities distributed in response to an existing or anticipated event. In the presence of a degraded communication infrastructure, such ad hoc sensor networks use the ability of each node to communicate directly with one or more other nodes in its physical vicinity, allowing message communication to take place via multihop spreading. Because these networks are self-organizing and decentralized, they can be highly reliable and degrade only slowly as individual nodes fail.

- *Network and information security.* Traditional security requirements also apply to disaster management, including, for example, the need for encryption of (some) communications, authentication of people and data sources, and ensuring the integrity of the infrastructure and the data. Indeed, a more "network-centric" approach to disaster management is vulnerable to attacks on the network, the information it carries, and the associated applications and processes. Because disaster management often involves collaboration among many organizations, often in ways that cannot be anticipated beforehand, it also imposes some unique require-

---

[1]The program was formerly known as Augmented Cognition. Further information on the program is available at http://www.darpa.mil/DSO/thrust/biosci/warfighter.htm.

ments: for example, flexible policies that can be changed on the fly and interoperability across different infrastructures. Moreover, the common security goal of data privacy may be less important, since the goal is often to disseminate data widely. Instead, security needs to ensure data integrity, authenticate the identity and role of users, and authorize access to different resources. The network and its applications must also be protected from denial-of-service attacks that attempt to overwhelm resources with invalid requests so that resources are unavailable for legitimate uses. Security research continues to make progress on these issues and has the potential to protect against malicious adversarial attacks as well as benign faults or loss of resources.

- *Biochemical sensors.* Identifying unknown substances can require a long chemical analysis using bulky and expensive equipment that is not suited to a rapidly evolving field situation. Rapid advances in electronics and biotechnology are beginning to make possible the creation of sensors capable of detecting specific toxins or deoxyribonucleic acid/ribonucleic acid (DNA/RNA) patterns. Effective use of such devices would help change chemical and biological agents from terrifying, invisible dangers to more manageable threats. Advanced biochemical sensors show the promise of low-cost field analysis tools. As the price of these tools decreases, they can be deployed widely in the environment to enable continuous in situ measurements. A toxic plume could be rapidly characterized as to its source and extent through sensors mounted on small, unmanned aerial vehicles that deploy automatically from rooftops with the first signs of the plume.

- *Infrastructure instrumentation.* Both disaster management and continuing social functions depend critically on the status of infrastructure for supplying water, power, communication, sewage disposal, and transportation. Currently, such systems have instrumentation at major critical points of failure—pumping stations, for example—but do not have distributed sensing throughout the network (e.g., to indicate broken pipes). Such sensing is critical for rapidly determining the overall status of the system after a disaster and effectively planning repair strategies as well as optimizing the use of the remaining assets while the repairs are being done.

- *Sensor systems.* One problem is how to track a person or event as it passes from the view of one subset of sensors to another. Multiple sensors also pose situational awareness and interpretation issues compounded by bandwidth limitations: what sensors should be given network priority and how multiple, geographically separated streams of information can be fused.

- *Integrated geographic information systems.* Long treated as a distinct specialty with a strong emphasis on detailed analysis and modeling, the

GIS field is merging into the mainstream of information technology. Base geospatial information and imagery are increasingly used as a context for real-time data in situational awareness and resource-management applications. At the same time, growing numbers of non-IT assets are becoming "location aware" by means of GPS and various wireless positioning techniques. Integration tools are bringing these disparate sources of geospatial information together to enable improved awareness and decision-making capabilities.

- *Portable unmanned aerial vehicles (UAVs) and (autonomous) robots.* The nature of disasters may prevent the normal collection of data by humans or require sensing capabilities beyond human capacity. Robots and unmanned systems deploying advanced sensors may be helpful in obtaining better and timelier information. Unmanned aerial, ground, and sea systems being used extensively by the Department of Defense could be adapted for use in disaster management. During the disaster prevention and preparation phases, unmanned vehicles could be routinely used as unobtrusive mobile sampling agents. During incident response and recovery, small aerial and sea vehicles could be used on demand by field teams to survey general damage and transportation infrastructure (especially bridges and seawalls). New underwater sensor payloads have been developed that should enable unmanned underwater and surface vehicles to monitor structures below the waterline. Small ground robots were introduced at the World Trade Center disaster to aid in searching through the rubble. Small UAVs were used in New Orleans and Mississippi in the Hurricane Katrina and Rita responses.

- *Epidemiological alerting systems.* Much more sophisticated uses of IT are becoming possible that offer the potential for earlier warnings and a finer degree of understanding of the spread of infection.[2] These systems also hold promise for mitigating the effects of bioterrorism and other emerging epidemiological threats, such as avian influenza.

## IMPROVED DECISION SUPPORT AND RESOURCE TRACKING AND ALLOCATION

- *Widely available collaboration software and file sharing.* Collaborative environments could enable information sharing to reduce duplication of

---

[2]For example, the National Electronic Disease Surveillance System (NEDSS) is an initiative that promotes the use of data and information system standards to advance the development of efficient, integrated, and interoperable surveillance systems at federal, state, and local levels; see http://www.cdc.gov/nedss/. See also ESSENCE, the Electronic Surveillance System for the Early Notification of Community-based Epidemics; available at http://www.geis.fhp.osd.mil/GEIS/SurveillanceActivities/ESSENCE/ESSENCE.asp.

effort. An example is distributed whiteboard applications for discussing and communicating decisions. Problems such as volume of traffic, number of sources, and data-consumption requirements must be better understood. A number of fundamental questions must still be answered, such as what happens when the volume of information on shared file sites scales up and the number of participants grows. Experience with consortia management systems also argues that further advances are needed. By allowing people to interact more easily and more often, better collaboration tools could also foster increased rapport and trust.

- *Intelligent adaptive planning.* Intelligent adaptive planning is the process of modifying a plan to adapt to changed circumstances. This type of planning breaks down the barriers between preparation and execution; it eliminates the dual problems of creating plans that are not followed and following plans that do not fit. There are several key requirements, each with potentially fruitful lines of research. (See Box 4.6 for further discussion.)

- *Computer-assisted decision-making tools.* Computer technology could be quite useful for decision making with tools like, for instance, online resource directories or "decision sentinels"—essentially a concept for monitoring crucial processes and activities. Decision sentinels attempt to raise warning flags when, for instance, things that are supposed to happen are not happening, when things that are not allowed for are indeed occurring, or when decision points are slipping. However, significant technical work is necessary in analyzing dependencies and assumptions in disaster plans—if these are understood, plan execution may be monitored accordingly, allowing for proactive warning of problems. Technology tools might also be useful in decisions involving dropping constraints, elevating conflicts for resolution, or switching to alternate plans.

- *Distributed emergency operation centers.* Distributed coordination, planning, and scheduling systems promise to provide tools for preparing disaster response plans and for coordinating and monitoring the execution in a more distributed manner.

- *Resource use modeling.* Detailed modeling and study of different disaster scenarios would give disaster response managers a much better idea of what resources might be needed, in what quantity, where and how best to use or deploy them, and so on—from items like potable water for survivors to issues as complex as the most useful deployment of satellites.

- *Exploring similarities between transportation networks and IT networks.* There is some fundamental abstract science common to moving "stuff" between nodes in networks, whether it is messages over communications links in a telecommunications network, signals over very-large-scale integration circuits in a computer, trucks over roads in the highway

> **BOX 4.6**
> **Intelligent Adaptive Planning**
>
> Intelligent adaptive planning breaks down the barriers between preparation and execution; it eliminates the dual problems of creating plans that are not followed and following plans that do not fit. The key aspect of intelligent adaptive planning is the ability to gather and assess information speedily from a huge variety of sources, including real-time sensors, and to package and disseminate results so that plans can be executed and changed. Intelligent adaptive execution is the process of modifying a plan to adapt to changed circumstances. There are several key requirements, each with potentially fruitful lines of research.
>
> One is avoiding *thrashing*—ineffectiveness caused when reacting too quickly to frequently changing situations sends too many resources toward ephemeral problems (e.g., food shipments that chase refugees in transit and never catch up with them). Another issue is containment of ripple effects. This is sometimes also referred to as *minimal disruption plan repair*—isolating parts of a plan so that the execution of the bulk of a plan can continue without fear that it will impinge on the part in need of repair. Minimal disruption plan repair is extremely important for two reasons. First, there are computational and organizational costs: the larger the span of activities that get reconsidered, the greater the time required and the greater the risk of expenses incurred from activities that become subject to revision or cancellation without completion. Second, there are human costs which, although difficult to measure, are nevertheless a substantial concern—when faced with constantly changing plans, people's morale and performance suffer. In addition, the loss of organizational credibility affects overall performance—people learn to expect changes and wait for their "real" orders to come rather than acting on the requests received. For all these reasons, thrash avoidance and minimum disruption plan repairs are essential issues.
>
> Three lines of research have the potential to contribute significantly to addressing these issues. The first applies not at the time of response but rather during plan preparation. *Dependency control* essentially augments what good military planners have instinctively done for centuries: attempt to increase the robustness of plans by trying to minimize the number of dependencies among subparts. Techniques for providing computational support for this effort range from analytic techniques that attempt to identify alternatives and change plans accordingly, to brute-force techniques, increasingly enabled by high-performance computing, that rapidly test alternative plans in simulation against thousands of alternative scenarios in order to see which holds up best. Use of these techniques helps ensure that the selected plan will work in the greatest number of situations and that problems, if encountered, will be confined to a restricted portion of the plan.
>
> The second line of defense against thrashing and disruption comes back to the topic of *uncertainty reasoning*. Conventional planning and scheduling techniques implicitly assume that the methods selected to perform tasks will succeed. In order to deal with the reality that things often go wrong, human planners (with some automated assistance) introduce contingency plans and decision points. These essentially attempt to predict where things might go awry, create fallback plans for addressing these problems, and insert points in the process where activities will halt to allow evaluation of the situation and a choice among the contingen-
>
> *continued*

> **BOX 4.6 Continued**
>
> cy plans. Although the need for this capability will never go away, uncertainty reasoning can make a great difference. Essentially, it allows a system to reason constantly about the probability of success of actions in progress and the degree of contribution that those actions will make to the overall outcome. This enables the system to put in play multiple alternative methods for ensuring a desired result—adding more actions if the risk increases, but also cancelling or redirecting some of them if the risk decreases. These techniques thus strike an appropriate balance between the risks and cost of failure versus the costs of duplicative "insurance" tasks.
>
> The third approach to avoiding thrashing and minimizing disruption investigates *stability-preserving planning*. Essentially, much of planning and scheduling is treated as an optimization problem. Criteria (often called a utility function) are established for defining a "good" schedule, and the various algorithms that the field has produced all represent different approaches toward trying to maximize that score. In stability-preserving planning, rather than planning from scratch, the system takes the pre-existing plan as input. Mechanisms for expressing the evaluation criteria are extended to permit description of the manner and extent to which a user desires retention of elements in the pre-existing plan to be weighted in the generation of the revised plan. This is, quite obviously, a highly desirable contribution to minimal-disruption plan repair.
>
> Combining dependency control, uncertainty reasoning, and stability-preserving planning techniques into intelligent adaptive planning systems will greatly increase the power of software systems to help organizations rapidly and effectively adapt to changing situations with minimum costs. Adaptive capability is most effective when coupled with proactive problem-identification systems based on techniques such as Plan Sentinels or mathematical progress indicators. These techniques point the way toward early-warning systems that provide a balance between maximizing the amount of time available to respond to problems with avoidance of overreaction to unlikely problems.
>
> Adaptation is easiest if the initial resource allocation planning is done well. New techniques offer the promise of smarter, more efficient algorithms that can produce better solutions to larger problems. Combined with better allocation algorithms, the concept of "best available understanding," in which the system fills in gaps by combining actual information with predictions and estimates, has the potential to avoid "garbage-in/garbage-out" difficulties by improving the likelihood that the problems fed to those improved algorithms best reflect the actual situation.
>
> Taken together, research in these areas has the promise of building systems 5 to 10 years from now that utilize resources far more effectively, recognize and adapt to difficulties and changed situations much earlier, and keep organizations efficient and credible in the course of doing so.

system, or people between cities via the air transportation system. Effective utilization of transport resources is increasingly tied to effective coordination of those resources using IT. Taking steps toward breaking down barriers between IT disciplines may stimulate out-of-the-box thinking about how IT research could be applied to solve problems with transportation logistics.

- *Logistics.* Flexible inventory and resource management systems would allow ad hoc and at-hand resources to be leveraged, while still offering the ability to provide information for an accounting of resource use, and could be based on standardized application of RFID tagging for asset discovery and tracking.

## GREATER ORGANIZATIONAL AGILITY FOR DISASTER MANAGEMENT

- *Computer mediated exercises.* Large-scale role playing exercises mediated by IT and involving multiple societal systems to help expose negative interactions and appropriate responses could advance response preparedness by allowing responders to gain "real-life" experience. This technology could also be used to incorporate disaster research into practice as well as provide an environment for testing such research.[3]
- *Flexible, embedded, and miniature projective visual displays.* The size, weight, and rigidity of existing visual displays (computer screens) can create usability problems under field conditions. An emerging generation of materials will permit digital display surfaces that can be flexed, folded, or molded, as circumstances require. Possible applications may include foldable/rollable map displays, fabric displays worn as parts of a uniform, and displays built into the surfaces of structures, vehicles, and roadways. Miniature projective displays offer a potential option when dexterity is an issue, such as when wearing gloves. Projective displays are small enough for a cell phone to project an image, such as a map or virtual keyboard, onto a flat surface. Heads-up displays, including virtual retinal displays using low-powered lasers, project information directly in the user's field of view.
- *Event replay tools.* As disaster management becomes more network-centric and storage costs decrease, the evolution of disasters (and responses) can be more easily monitored and archived. These capabilities allow ongoing replays of what just happened and post facto analysis and documentation of lessons learned.
- *Online repositories of lessons learned.* Lessons learned in one region of the country or in one disaster management domain could be more rapidly and broadly disseminated if a well-known searchable repository was set up. The federal government's Lessons Learned Information Sharing (www.llis.gov) Web site is a first step in this direction.

---

[3] An example of this technology applied in practice was presented to the committee. See Synthetic Environment for Analysis and Simulation (SEAS) at http://www.mgmt.purdue.edu/centers/perc/html/index.htm.

- *Integrated ad hoc data-collection tools (blogs/wikis).* A variety of Web tools are available that allow a collaborative and distributed assessment of a situation. These Web tools can potentially be viewed and augmented by disaster management personnel and the public at and away from the scene.
- *Continuous learning tools.* Web-based training can allow first responders to be proficient with new information technology. The training can be customized to individual first responders and available on demand at their convenience. Adaptive training technology continuously monitors understanding of and facility with the new knowledge, providing additional training depth or skipping material as needed.
- *Computer-assisted disaster simulation training.* Computer-assisted simulations offer task-oriented training based on real-world situations. They offer a persistent "world" for continuous replication, verification, validation, uncertainty quantification, and margins-of-error estimation. They also provide an outcome-based learning experience that can lessen the effect of affiliation goals, which may drive decision making over situation needs in early stages of disaster response. An example of the possibilities in this field is the Virtual Terrorism Response Academy developed by Dartmouth's Interactive Media Laboratory.[4]
- *Distributed, scalable, survivable data logging.* With advances in sensor technology, it is increasingly feasible to capture data that might be very useful in disaster management (e.g., levee saturation levels, building or structural stresses, and so on). Major challenges, however, involve things like integrating such data into useful tools, making sure that such data survive a given disaster, and archiving the data in such a way that they can be learned from for future disasters.
- *Dynamic capability profiling and credentialing.* Integrated systems for registering and credentialing people with skills required for disaster response could improve the effectiveness and efficiency of response and recovery operations.
- *Digital identity.* Disasters frequently require individuals and agencies to forge new working relationships with others they do not know. A major disaster may involve responders from all over the nation or even the world. The lack of trusted mechanisms for determining the identity, capabilities, and privileges of new response partners can delay or even deadlock critical information-sharing activities. Public key technology and other techniques for digital authentication and authorization are rapidly becoming irreplaceable elements of the emergency information infrastructure.

---

[4] See http://iml.dartmouth.edu/vtra/.

- *Dynamic authority mapping.* Web-based command and authority charts to determine who is responsible for decisions as a disaster develops could improve understanding of how organizational structures map onto communication structure requirements and could improve the efficiency of response and recovery operations.

## BETTER ENGAGEMENT OF THE PUBLIC

- *Multimodal public notification.* Multiple modes of communication (e.g., e-mail, Web pages, and cell phone) and multiple types of information (e.g., text, audio, and video) can all be employed to mobilize volunteers and private voluntary organizations and to provide more useful information to the public. In addition to delivering messages in multiple languages, such systems could incorporate visual (e.g., sign language), sound, and tactile communications to reach different special-needs groups.
- *Multimodal public reporting systems.* The best-known example of a public reporting system is the 911 emergency calling system. It has proved highly effective as an efficient means for the public to notify public safety officials of a situation requiring their attention. Enhanced-911 (or E911) extends the 911 system to wireless (cellular) devices. While valuable, the 911 system is exclusively voice-communications-based. Technology is available and could be applied to extending public reporting systems beyond voice to include text, data, image, video, and so on.
- *Enhanced two-way communications with the public.* Technology capabilities are now making it possible to send alerts and warning notifications and instructions to specific geographic areas or entire regions to a range of devices, including cell phones, pagers, computers (e-mail), and wireless PDAs. (One current capability of this sort is so-called reverse 911, which provides an automated way of contacting citizens at risk.) This is an important first step in improving two-way communication with the public. Further advances may allow notices to be finely targeted, with messages tailored to individuals on the basis of the type of risk, proximity to the hazard, and other important factors. In addition, tools for collecting voice (e.g., 911), text (e.g., e-mail and SMS), and image (e.g., picture phone) may allow systematic incorporation of data reported by the public into the process of developing a broad-based view of a disaster. Perhaps more importantly, these systems may eventually enable true risk communications as an interactive process of information exchange between officials and the public.[5]

---

[5]National Research Council, *Improving Risk Communication*, National Academy Press, Washington, D.C., 1989. This report draws a careful distinction between risk messages and the risk communications process.

- *Validated online information sources.* During a disaster, the Internet allows information that is rapidly evolving, voluminous (e.g., missing-persons lists), or complex (e.g., a tiered evacuation) to be disseminated and presented efficiently. How can sources of such information be validated?
- *Volunteer mobilization systems.* Directories would make it easier to identify and use volunteers, organizations, and commercial resources when needed. Such directories could be connected to communication systems so that these various resources can be automatically tracked down and called on as needed.
- *Game technology.* Simulated disaster management training environments based on gaming technology are starting to be applied to educate the public. Further developments in this area may enable the public to gain a much greater understanding of what to expect in disaster situations and provide insights on how they can best prepare for and participate in disaster response.

## ENHANCED INFRASTRUCTURE SURVIVABILITY AND CONTINUITY OF SOCIETAL FUNCTIONS

- *Mobile power generators.* Extended widespread disasters stress the power grid and backup power sources. Efficient, compact, and easily deployed power sources can play a central role in the restoration of communication, the provision of medical services, and so on. While generators are pre-deployed at many facilities, a system for tracking critical unmet power needs and distributing power resources can facilitate more rapid recovery.
- *Communications redundancy.* Different communication infrastructures are susceptible to different types of failures. Having access to different communication technologies increases the chances of being able to communicate during or after a disaster. Certain wireless technologies (e.g., mesh networking) are especially attractive, since they require little or no infrastructure, can be power efficient, and can be used to extend the reach of functioning communication infrastructure.
- *Embedded sensors for non-destructive asset evaluation.* One of the challenges after disasters is determining what parts of the physical infrastructure (e.g., buildings, bridges) are safe to use. This often requires a physical inspection by trained specialists. Embedded sensors can speed up the evaluation of physical assets, thus speeding up the recovery process.
- *Automated monetary disbursement.* A major challenge in widespread disasters is providing people with financial resources to cope with immediate needs. Systems are needed for entering claims, authenticating re-

cipients, and tracking transactions. Web-based tools can enable faster and more secure claims processing.

• *Risk management tools with uncertainty modeling.* Advanced modeling that incorporates experience with risk assessment, including cost-benefit metrics, of societal-scale systems before, during, and after disasters could improve mitigation and preparedness investment decisions.

• *Resilient materials and structures and deployable infrastructure better adapted for the built environment.* Building construction has been modified and existing buildings retrofitted with technology improvements for better earthquake survivability. Similar advances are possible to improve the resilience of communications infrastructure; better meet communications needs inside buildings or other enclosed spaces, including damaged structures; and provide advanced sensing and instrumentation of structures. Advances have implications for preparedness, response, and recovery. Hardened repeaters in buildings, low-frequency radios, and "bread crumb" repeaters deployed by first responders as they advance through a structure are some of the possibilities.

• *Replicated and secure medical databases.* Patient records should be remotely available to authorized personnel. Patients can be tracked and medical records can automatically follow them within improvised medical facilities using RFID or other related technologies. Such capabilities raise obvious privacy issues, including whether and how to relax privacy constraints in a disaster situation.

• *Person tracking and reporting.* Networking technology can allow family members a better means of connecting with one another—for example, by enabling standardized finder databases or database schema for collecting and disseminating people's whereabouts.

# Appendixes

# A

# Illustrative Fictional Narratives of IT Use in Disaster Management

The Committee on Using Information Technology to Enhance Disaster Management developed three narratives describing fictional disasters illustrative of the current and potential use of information technology (IT) in disaster management practice. The purpose is to provide a realistic context for understanding and analyzing potential goals and capabilities for the use of IT to improve disaster management now and in the future. The committee constructed narratives representing different types of disasters with different proximate causes, varying lengths of warning, different geographic scope, and different levels of responder perspective (local/tactical, regional/strategic and tactical, national/strategic) in order to uncover both common and unique problems that they present. The narratives draw on the experiences of disaster management practitioners and researchers and highlight specific aspects of technology use. Each scenario shows elements of IT use across the goal areas identified in the report, though a scenario may naturally highlight the importance of some goals more than others. Some descriptions may seem more compelling because they are closer to the direct human impact of a disaster, but gaining an understanding of and improving on disaster management practice at all levels are critical to saving lives and reducing economic impacts. Taken together, the scenarios build a picture of the many moving parts that must be coordinated for an effective response.

For each scenario, an introductory paragraph describes the setting and the focus for analysis of IT use. Each scenario concludes with a summary analysis that highlights some of the challenges and opportunities for potentially improving IT capabilities. The committee emphasizes that, although they are set in specific contexts, these scenarios are fictional and are not intended to reflect in detail the specific response capabilities of particular jurisdictions or agencies.

## INITIAL RESPONSE TO A CHEMICAL ATTACK IN THE WASHINGTON, D.C., METRO

The first fictional narrative describes the exchange between responders during a chemical attack in the Washington, D.C., Metro system. It presents the use of information and communications technology from the perspective of local and tactical first responders during the onset and immediate aftermath of a terrorist attack. The focus of this scenario is on the use of communications and information infrastructure by the police, emergency medical personnel, other responding public safety personnel, and more generally by anyone (including individual private citizens) who can act as first responders. This scenario highlights a number of uses to which IT is currently put in responding to situations that occur with no warning and rapid onset. The response described here is entirely tactical, and the technology used is what has been predeployed and is on hand or can rapidly—in seconds and minutes—be brought to bear. The alert reader will identify a number of technologies used: video, chemical sensors, public safety radio, operational control systems (Metro system), commercial cellular telephone, commercial land telephone, and text messaging during the first few moments. Metro dispatch; firefighting, emergency medical, and hazardous materials personnel; Metro police; emergency managers; and Federal Bureau of Investigation (FBI) units are all rapidly engaged and seeking information.

*7:46:00 a.m.*
**Metro Dispatch** *(Mary Williams—33 years old, 5 years' experience as a dispatcher):*
 *"Unit 332—Report of person down—lower level, Metro Center."*
**Unit 332** *(John Bison—26 years old, former military):*
 *"I'm close. I'll respond." John finishes his coffee, drops the cup in the trash, and begins to walk quickly to the stairway leading to the lower level.*
**Metro Dispatch:**
 *"10-4."*

APPENDIX A						137

*7:46:30 a.m.*
**Metro Dispatch:**
"Unit 332—We are getting multiple reports of multiple people down. Showing Blue line due in the next minute. Orange and Red line trains due into the station within the next 2 minutes. Rolling paramedics and additional units . . . all units responding to Metro Center lower level keep the air clear until at scene. Live feed video shows people running in all directions."
**Unit 332:**
"10-4—10 seconds out. Lots of people screaming and running toward me."
**Unit 337** *(Billy Boyd, 53 years old, 29 years with Transit Police, 63 days from retirement):*
"Unit 338 and I are close and responding."
**Metro Dispatch:**
"Units 300S and 300L, are you monitoring this?" *(Unit 300S is the sergeant and 300L is the lieutenant for the shift.)*
**Unit 300S:**
"Unit 300L is with me and we are 5 minutes out."

*7:46:40 a.m.*
**Metro Dispatch:**
"Emergency—All units' chemical sensors going off, lower level Metro Center. Unit 332, do you copy? Unit 332? Unit 332? Unit 337 or 338, do you copy?"
**Unit 300L:**
"Metro Dispatch, freeze all inbound trains."
**Metro Dispatch:**
"Working on it. I'm unable to raise Units 332, 337, and 338."
**Unit 340:**
"I'm en route. I'll check on the units—what was their last location?"
**Metro Dispatch:**
"Unit 340, that's a negative—we have chemical sensors going off in the lower level Metro Center. I need you to begin full evacuation procedures for Metro Center. The last known locations of 332, 337, and 338 were. . . . Unit 300S—I'm getting emergency signals from 332's and 337's radios."
**Unit 300S:**
"10-4—Let's get everyone we can evacuate without entering and get the folks with Level A suits in to help the others. I'll need an ETA on the chemical team."

*7:47:15 a.m.*
**Metro Dispatch:**
"Unit 300L—Orange and Red lines both frozen 1 minute out. It looks like the Blue line arrived and opened doors before realizing what was going on. . . . They appear to be stopped at the loading area."
**Unit 300L:**
"Command Post will be on G St. between 11th and 12th. I need the next four units to cover the four entrances into Metro Center. Have the hazmat team meet me at the command post. I want all units in full MOP gear with their masks."
**Metro Dispatch:**
"10-4. Fire should be pulling up now. Hazmat is 2 minutes out. We have alerted area hospitals and requested additional ambulances to your location. The chief has ordered the EOC [Emergency Operations Center] opened."
**Unit 300L:**
"10-4. D.C. Metro is with us now, but we need as many more units as you can find in the area. We are trying to contain those exiting the Center, but are overwhelmed. Several have exited showing signs of chemical exposure. Possibility of off-gassing from those leaving the area—can the chief advise on procedure for those refusing to stay in the area?"
**Metro Dispatch:**
"10-4. Checking."

*7:57:00 a.m.*
**Unit 300L:**
"Fire command is establishing the decontamination center at SW corner of 11th and G St. Have all units direct those leaving Metro Center to that location. Everyone . . . I repeat . . . everyone exiting Metro Center must go through the decontamination center and be cleared by hazmat."
**Metro Dispatch:**
"10-4—FBI is asking for your location and EOC would like an update when you have a moment. Line into the EOC is 444-1234."
**Unit 300L:**
"I'm standing at the corner of G St.—have FBI respond to this location. How about additional units? Can we get some help for Capital, D.C. Metro?"

As 300L makes this request, he is attempting to dial the EOC on his cell phone, a commercial network that is overloaded with traffic.

APPENDIX A

**Metro Dispatch:**
"Lots of units from different agencies are responding to your location, but many are reporting being stuck in gridlock."
**Unit 300L:**
"Dispatch, can you call the EOC and have them come up on TAC Frequency 1—my cell is worthless down here right now—I'll brief them over the radio."
**Metro Dispatch:**
"10-4—Fire is requesting additional security at the decontamination area. They also said to let you know they are about 10 minutes from getting into Metro Center with Level A suits. They are suggesting that any of our personnel on the inner perimeter get into their MOP gear if they aren't already."

Meanwhile at the EOC
**Unit 300C** (Unit 300C is the captain):
"Where is 300L? How come he has not called in?"
**Sally** (staff member in EOC):
"Sir, communications is asking that we come up on TAC frequency 1[1]—300L apparently can't get out on his cell."
**Unit 300C:**
"Thanks. Flip us over would you . . . and can you check on how we are doing evacuating those other trains?"
**Sally:**
"I'm on it, boss."
**Unit 300C:**
"300L, are you on TAC 1?"
**Unit 300L:**
"That's affirm, boss."
**Unit 300C:**
"How's it going out there—what can we get you?"
**Unit 300L:**
"It's a mess right now—not quite sure where everyone is . . . have not had a chance to do a roll call. Numerous people down—we have at least three units inside Metro Center that are not responding. Fire has decontamination set up and are washing down, but we can't control everyone coming out and direct them to the decon area—I'm afraid we are going to get some off-gassing exposures. Here's what I need. As many people with full MOP suits as you can get down here. I have

---

[1]TAC frequency 1: Land mobile radio (LMR) 800 MHz car-to-car frequency.

D.C. Metro with me, but I still need reps from FBI, fire, Red Cross, and medical services so we can coordinate our response . . . it looks like this is going to be real ugly. . . . The Blue line pulled up to the platform and has not moved . . . it had to be a full train at this time. . . . Hazmat is about 5 minutes from getting into the station in Level A suits. Another thing—can you get SWAT fired up just in case someone trips over a suspect?"

**Unit 300C:**
"We will do our best to get it to you—let us know if you need anything else, and be careful."

**Unit 300L:**
"Thanks."

**Sally:**
"Sir, we are getting a report of another attack at L'Enfant Plaza—sounds like two trains in the station at the time, chemical sensors going off, and numerous people down."

**Unit 300C:**
"Can you give me the video feed from there?"

**Sally:**
"I can give you the picture, but I can't give you control of the cameras—ops says they need it right now. Picture should be coming up momentarily."

As the video of L'Enfant Plaza comes up, everyone stands motionless. As the camera moves, they see fellow Transit Authority workers motionless. They are dead—of that there is no doubt—along with hundreds of others.

**Unit 300C** (looking over to the Fire Battalion Chief):
"Is there anything you can do?"

**Fire Battalion Chief:**
"Not for them . . . most of my resources are at Metro Center. Capital hazmat is en route, but . . . we can't even triage the people coming out and when we do the hospitals have to re-evaluate, as we can't pass the vital info to them. How about you?"

**Unit 300C:**
"We are bringing in a number of people from Metro, Capital, and Arlington Police—so staffing should be OK in a little—but until we can get them in there we are stuck. Our comms are a mess—the two incidents are on the same frequency and we have not been able to break one off yet. Our patches are working with the other police agencies, but the feds don't have the capabilities or won't let us patch with them. Video is OK, but I don't want anyone in the field seeing it. Cell is overloaded and connections are intermittent. I used my GETS card on

APPENDIX A                                                                    141

> *the hard-line a moment ago and that worked fine. I understand that text messaging is working on cells, but only our command staff has them. One of the sergeants used his own personal computer to log in and send an e-mail over an unsecured 802 wireless network—the Mobile Computer Terminals all have 802.x wireless capability but it's turned off. Lastly, we are still having trouble tracking the people and resources responding into the area. We really need to get a handle on that. . . .*

An analysis of this scenario reveals a number of notable problems with and limitations of IT use. These include radio-frequency conflicts; the lack of means for communicating with victims (using either voice or text) or for obtaining information about them (through automatically transmitted sensor data from their devices); impaired video capability; limits of commercial cellular infrastructure; lack of access to redundant or alternate technologies (e.g., text messaging, wireless Internet); inability to disseminate and share information; reduced timeliness and limitations of sensor data; absence of dissemination of sensor data to responders, hazmat units, and hospitals; and inability to distinguish different information needs of various responders. The ability to interpret the data rapidly (either by getting it to the appropriate expert or through automated analysis) and then to issue appropriate and timely instructions to responders and affected individuals based on the analysis of the data is clearly critical. Some of these problems and limitations could be addressed with existing technology; others require technology advances.

A striking aspect of the response described is the need to gather and disseminate information rapidly in order to create sufficient situational awareness for making effective response decisions in a rapidly evolving situation. For instance, chemical sensors started going off only 40 seconds after the initial report of a casualty—and yet this was already too late to avoid dispatching responders when they could no longer provide effective aid and to keep doors from opening on entering trains. Detection of harmful agents might, for instance, trigger automated response systems to halt or reroute the entry of additional trains into the area. Automated analysis of sensor data could also alert the dispatcher to take appropriate measures to protect and direct responders—not just to avoid putting them in harm's way but also to make sure that they could be effective if they were deployed. Other possibilities include automated control of heating, ventilation, and air conditioning systems based on previous plume modeling. Understanding what can be automated and striking a balance between what should be automated and what requires human decision makers are major challenges revealed by an analysis of this scenario. Finally, the scenario does not consider how the increasingly sophisticated mobile consumer devices that Metro employees are likely to carry could be used to provide better information about the unfolding incident.

## THE UNFOLDING RESPONSE TO A HURRICANE

The second fictional narrative describes the development and progression of a major hurricane in the Gulf of Mexico and the corresponding disaster management response that is possible now with the current level of technology. It focuses on the use of information and communications technology from the perspective of the regional Emergency Operations Center. The scenario shows how emergency managers use IT during the pre-incident onset phase, through landfall, to the early stages of post-incident recovery. This scenario highlights a number of uses to which IT is currently put when responding to situations with considerable warning and slow onset. Situational awareness and a common operating picture are a particular focus. The substantial lead time and detailed knowledge now available prior to the onset of a hurricane are possible owing to great advances in IT and other scientific disciplines.

The response described here is a mix of tactical and strategic operations. Technologies highlighted in this scenario include computer modeling, simulation, weather sensors, satellites, WiFi and mobile networking, traffic cameras, and unmanned vehicles with onboard communications. For this scenario, information and communications flows are centered largely on the state Emergency Operations Center, with raw data flowing in, being processed, and analyzed; actions being formulated; and instructions flowing back out.

*A Category 4 hurricane is churning in the Gulf of Mexico, building strength. It is projected to be the worst storm surge ever experienced on the Florida west coast. The hurricane begins to make its way toward Cedar Key, a sparsely populated area near Ocala, with expected landfall on Saturday at midnight. Emergency managers develop an Incident Action Plan (IAP) for the response using previous plans and the results of a distributed simulation exercise. IAPs are specific to an event. At this point in an event, the communities' and state's emergency management plans form the framework of response. This approach allows each agency to see its role in the plan and identify possible bottlenecks. Alternative plans are developed for possible storm deviations.*

*The state emergency manager activates the plan for Cedar Key based on computerized weather models showing the highest probability of landfall occurring there. Local and state response teams (e.g., search and rescue, law enforcement) are prepositioned near Cedar Key, and federal teams are prepositioned along the path of the hurricane, which is project-*

APPENDIX A

ed to cut across Florida and then up through the interior of Georgia and Tennessee.

The hurricane begins to be felt around 8 p.m. The entire area is under a hurricane warning issued by the National Weather Service. The Weather Service, taking advantage of a network of wind and wave sensors erected along the west coast, begins to see in real time that the port city of Tampa is unexpectedly bearing the brunt of the storm, indicating that the storm is not heading north but has turned slightly east, mirroring the behavior of Hurricane Charley. The state EOC dynamically switches to an alternative IAP previously developed for this area, using simulations and drawing on a cache of plans from previous responses.

Based on the alternative plan, the state EOC generates modifications to the existing response plan, notifying all critical agencies (i.e., those designated in the plans as key agencies for interagency emergency communications and operations), rerouting Coast Guard flyovers and prioritizing night flights. The broadcast media interrupts programming to saturate the airwaves with hurricane warnings, and the state communications office sets up a 511 number for information. Law enforcement officers are dispatched to close down entertainment areas, alert the areas most at risk, inform them of how to respond, and survey the status of voluntary evacuations, identifying specific high-risk targets such as nursing homes that have not been evacuated. The Department of Transportation (DOT), another designated critical agency, brings in crews and orange cones to help route traffic out of the popular bayside night club district. Traffic cameras show real-time traffic flows along designated evacuation routes, and DOT forwards assessments to the media for broadcast to the public.

The hurricane makes landfall at midnight, but stalls unexpectedly. The Weather Service projects 18 inches of rain in the Tampa Bay region by morning. The rainfall, coupled with the storm surge, could topple the levees surrounding toxic phosphate mine tailings. The state emergency manager calls for the deployment of pre-identified assets for the large-scale testing and monitoring of water quality and designated environmental mitigation teams. EOC staff work through the night with computer programs projecting damage to key areas, residential areas, and infrastructure; these become rapid reconnaissance plans. Intermittent reports from pre-positioned recon teams and law enforcement provide some indication of damage, but the winds and rain must die down before the Coast Guard can fly or reconnaissance teams can venture out. The governor and president are briefed about potential ecological and economic damage, based on models and the current storm path.

The hurricane finally passes in the early morning hours, and the wind and rain begin to subside. Electric power and cellular coverage were lost early on due to wind damage from the leading edge of the storm, and satellite phones are waiting for the skies to clear. Recon and response teams are set to mobilize at 4 a.m. Teams have refined their plans throughout the night, generating search grids for the area, reviewing land use maps, and assigning teams and assets. As they begin to deploy to the field, they find it difficult to reach designated areas due to extensive damage to transportation infrastructure. Limited ability to communicate difficulties and receive updated damage information further hinders progress.

A large neighborhood near downtown is flooded, and survivors are on the roofs, while a major senior and assisted-living development in the suburbs previously identified as not having been evacuated is also damaged and flooded. Rescue teams deploy miniature unmanned helicopters to geolocate survivors on roofs, remotely talk with them, and also look around roofs and attics for signs of survivors trapped inside. The Coast Guard sends manned helicopters to the suburbs to lift survivors from the rooftops, while the rescue teams talk directly with the pilots and provide prioritization and data on rooftop conditions from a portable anemometer. The majority of the rescue teams focus on determining who in the urban area needs immediate evacuation. When the rescuers find a group of looters, they alert the police, who in turn dispatch the nearest law enforcement personnel. Boats are tasked to help the elderly and handle medical priorities, but routing them to appropriate care facilities is hampered by damage infrastructure making it impossible to regularly communicate information about the operational status of medical facilities.

As operations transition slowly from response to recovery, law enforcement agencies attempt to separate curious bystanders from returning authorized responders, residents, and repair crews. But, due to limited means to make such distinctions, they tend to deny access to some appropriate personnel—slowing the repair and recovery process. Utility providers (e.g., electrical, water, communications) gather information about damage and determine how to re-establish services. Portable radio broadcast towers are erected and by 7 a.m., news and response directions are being transmitted to the public in the affected areas. While the EOC has a general picture of where resources have been dispatched and has pieced together an overall picture of damage from information coming from field operations, detailed situational awareness continues to be elusive.

Information and communications technology is obviously critical to managing the widely distributed and massive operations described in this scenario. Indeed, IT is used effectively from identification of the emerging situation through to the recovery phase. Lives are saved, damaged reduced, and steps toward recovery made rapidly by extensive use of IT. Yet, there are many opportunities to extend and improve the use of IT. A combination of existing technology, further technology advances,

APPENDIX A

and organizational innovations should be able to extend the excellent situational awareness that was available prior to onset into the response phase, and beyond to the recovery phase. Interagency communication and coordination are especially critical in this scenario, and the planning operations described stress this point.

However, the single global situational awareness developed at the EOC is inadequate for developing the task-specific situational awareness needed by different participants. For instance, a search-and-rescue team incident commander should see where search teams are, what they have searched, and what the results are. He or she would have the ability to report looting or unauthorized entry to law enforcement, using authorized access to directly input the information into the system managing the operational picture. The medical community should also see the progress and projections of injuries, heatstroke, local capacity for medical care, and so on—and be able to send information about its situation to other operational units for (automatic) inclusion into their operational picture. The EOC should receive continuously updated information that can then be used to revise its operational picture on an ongoing basis using adaptive planning tools.

While transportation officials are able to identify traffic congestion using cameras and to communicate problems that have already developed, use of traffic sensors that measure speed and volume of traffic could be incorporated with information about infrastructure damage and other situational variables into decision support systems to recommend alternate routes proactively before the transportation system bogs down. Civilian volunteers at homes and shelters, fire departments, and police stations could be pre-equipped with WiFi network capability and designated as an integral information source in emergency plans. Civilians in the damaged areas with functioning IT systems (e.g., cell phones) could have a channel for sending pictures and other information from areas to a designated repository that can be used to build situational awareness.

With technology available and on the horizon, it is possible to imagine rescue teams, as they drive into designated areas, receiving constant updates from DOT reconnaissance teams on road conditions and damage. A roadmap on a responder laptop could glow with red overlays on known or suspected damaged bridges. Navigation software could attempt to create directions for each of the teams that would get them to their assigned grid. Driving in, a team member could use infrared and laser cameras to videorecord the geolocated damage. As a team passed a WiFi hot spot, the laptop could download new information from flyovers, and the Weather Service could transmit video to the incident commander on a priority basis and then to the EOC. Decision support tools could begin combing emerging information, correlating and analyzing who needs what kind of help and what the transportation conditions are. They could also identify areas where no information is available and target quick reconnaissance teams to check out those blind spots. As new and better information becomes available, adaptive planning tools could be used to change response activities to reflect the evolving situation. Visualization software could present task-specific views of the situation for each emergency support function, and provide an integrated view of the overall situation. Law enforcement could have access to centralized databases, with pre-identified authorization for efficient and accurate screening of affected areas.

## NATIONAL RESPONSE TO AN EARTHQUAKE IN THE SAN FRANCISCO BAY AREA

The third fictional narrative describes the occurrence of a major earthquake in the San Francisco Bay area and the resulting response. The focus is on the use of information and communications technology from the perspective of the federal-level emergency manager as part of a national response. In this case onset is rapid, but such an incident has been anticipated, and detailed plans exist for responding to it. The blizzard of acronyms blanketing this scenario is a clue to the complexity involved in coordinating a response at this level and the degree and sophistication of interagency interaction required. Resource allocation, logistics, tracking, and mobilization are of particular concern in this scenario. Planning triggers, such as the declaration of an Incident of National Significance, also affect what IT resources are applied and when. Most of the technology employed in the two fictional scenarios above would also be applied in this scenario. However, technologies (and issues) highlighted here are those focused primarily on strategic response: for example, collaboration tools, modeling, information analysis and filtering tools, logistics and tracking, system scalability, and geographic information systems (GIS).

*At 4:03 a.m. on a Monday in March, a magnitude 7.3 earthquake occurs on the Hayward fault, with an epicenter very near the University of California, Berkeley. The Hayward fault event is not unanticipated and has been studied by scientists and emergency managers for years. However, in the darkness and confusion surrounding the early hours, it is difficult to get an accurate picture about what has just occurred.*

*State Office of Emergency Management (OEM) officials in Sacramento and Federal Emergency Management Agency (FEMA) and Department of Homeland Security (DHS) officials in Washington, D.C., get their initial situational awareness from prior modeling results: the Association of Bay Area Governments, FEMA, and the state of California have developed loss-estimation models for this scenario. Initial estimates are that more than 100,000 dwelling units are severely damaged in the Bay area, primarily in Alameda and San Francisco counties. The models predict that almost 400,000 people will be displaced and will need food and water, and 110,000 will need emergency shelter. Oakland and San Francisco airports and the Port of Oakland are expected to be closed, as are major interstate highways (I-80, I-580, I-880, Highway 13, Highway 101) and bridges (Bay Bridge, Dumbarton Bridge, Richmond–San Rafael Bridge). Models developed by the University of California, Berkeley, predict severe damage to that institution. Significant casualties are also expected.*

*The information from the models is enough for the secretary of DHS to declare an Incident of National Significance, to activate the Catastrophic Incident Annex of the National Response Plan, and to deploy specialized resources. Resources deployed include Rapid Needs Assessment Teams, FEMA Advance Teams, Urban Search and Rescue Teams (USART), National Disaster Medical System (NDMS) teams, and Disaster Mortuary Teams. The mobilization of government and Red Cross resources is also begun.*

*The common operating picture and situational awareness problems begin almost immediately. Due to bridge and highway failures and communications problems, the Bay area has been transformed into four or five disconnected "islands," complicating coordinated response efforts. Initial communications with FEMA Region Nine are restricted to satellite phone. Although Travis Air Force Base is open, it is not clear to emergency managers where the greatest needs are and how to bring resources to bear to meet them.*

*Rain and fog impede overflights and generation of satellite imagery in the hours immediately following the incident. Scattered reports from citizens, media, and responders from the area are conflicting and vague. For example, reports have been received that the Bay Bridge is "down," but no one in Washington, D.C., or Sacramento knows if that means that the bridge has collapsed, a section has collapsed (as occurred during the Loma Prieta earthquake), or an on-ramp is closed. Similarly, unconfirmed reports have been received from the media that there are "thousands" of casualties at the University of California, Berkeley, and that people are trapped in collapsed buildings in Oakland.*

*In Washington, D.C., the DHS Homeland Security Operations Center (HSOC), the FEMA National Response Coordination Center (NRCC), and the White House Homeland Security Council (HSC) have delivered conflicting initial assessments to the president: FEMA NRCC's assessment is based on HAZUS model output; DHS HSOC's assessment is based on initial reports from the media, NORTHCOM, and the State of California OEM; HSC's assessment is based on conversations with the governor of*

California. The HSOC and the NRCC were operationally, but not physically, combined after Hurricane Katrina into the National Operations Center (NOC). The Berkeley earthquake is demonstrating that the two operations centers are still operating independently. The secretary of DHS has convened his interagency Incident Advisory Council (IAC) and is receiving information directly from senior officials in other cabinet offices. For instance, health and medical information, including the status of medical assets in the region, is processed through the Department of Health and Human Services (DHHS) EOC. The president's homeland security advisor has convened the Domestic Readiness Group (DRG) to provide policy advice to the president and to the secretary of DHS.

Even after communications are established, the situational awareness problems persist. The Homeland Security Information Network (HSIN), developed to provide uniform, richer, information-sharing capabilities and collaborative tools, has been overwhelmed. Information available through HSIN is not quality-checked and is inconsistent, inaccurate, and incomplete. The information available is changing rapidly, and it is difficult to determine the source or timeliness of the information. In particular, GIS imagery is available on HSIN from several sources (State of California, U.S. Geological Survey, FEMA), and different images convey different information. No process exists for updating and date-stamping situational information as reports from the field ground-truth initial estimates produced by models, media, and rapid needs assessment teams. In particular, the status of critical infrastructure (e.g., road closures, water availability, sanitation) is difficult to determine, since reports are using non-standard, non-technical, and inconsistent language.

First responders and infrastructure managers, engaged in critical life safety and lifeline repair tasks, cannot provide consistent or complete information. Managers in the NOC are suffering from information overload and are beginning to narrow their focus to one or two sources of situational information (often CNN and one source from the field), a typical human response. HSOC is relying on reports from the state OEM, and the NRCC is relying on the FEMA Advance Emergency Response Team. The pre-designated Federal Coordinating Officer has been briefed by the NRCC and has been deployed to Travis, as have the pre-designated Primary Federal Official (after briefings from DHS HSOC) and the pre-designated Defense Coordinating Officer (after briefings from NORTHCOM).

The White House DRG is collecting information from all available sources. On Day 2, as a Joint Field Office (JFO) and unified federal/state command is established in Sacramento, it is apparent that the operating picture of the president, the secretary of DHS, the governor, the PFO, the FCO, and responders on the ground in the Bay area is not yet a common one. It is also clear that even where there is a common understanding of *situational information, substantial differences exist in meaning attributed to that information depending upon the background, experience, and organizational responsibilities of the manager/decision maker.*

*For FEMA, the American Red Cross, NORTHCOM, and the U.S. Army Corps of Engineers, this is the largest mobilization and deployment of people and resources since Hurricane Katrina. The Corps has activated its contracts with vendors and has ordered hundreds of truckloads of bottled water and ice. New radio-frequency identification (RFID) systems will track the relief commodities from the point of manufacture, to federal warehouse sites, and to state and local distribution sites. Attempts will be made by the Corps to ensure that DHS and FEMA have access to this logistics information to enable them to assess the gap between needs and supply. FEMA has activated pre-scripted mission assignments with the National Communication System (for cell on wheels and switches on wheels to restore cell phone communication systems) and with DOD (for Navy vessels for shelter and command support, for Army and Air Force logistics, and for troops to assist the California National Guard).*

*The American Red Cross is mobilizing 50,000 volunteers using its Disaster Services Human Resource system, and FEMA has mobilized and deployed its disaster reserve corps. NORTHCOM has established Joint Task Force California to manage the influx of Department of Defense resources. FEMA has activated its National Emergency Management Information System (NEMIS) and its call centers to create the ability to register disaster victims for individual assistance. NEMIS has been dramatically expanded since it collapsed during Hurricane Katrina, but it has not been tested. As Day 3 progresses, DHS, FEMA, and White House managers continue struggling to determine both what the situation is on the ground and to assess the adequacy of the response resources on scene and resources in the pipeline.*

The scenario above exposes the complexity involved in coordinating numerous agencies at all levels, with often widely overlapping and conflicting responsibilities but different perspectives. Those perspectives as well as organizational culture and instincts affect how information is interpreted, what information is trusted, and even where information is sought. This has profound implications for decision support tools, logistics, and resource allocation systems. One specific implication is how IT systems deal with incomplete or uncertain information and how such information is presented to disaster managers. Another is integration of and coordination with civilians and ad hoc groups in this environment. How can the volunteer groups that stand up be integrated and coordinated? The scenario illustrates the extent to which hierarchical command and control are assumed in organizational structures and how that is reflected in the implementation of IT systems. Introducing distributed networks and applications—and the organizational and cultural innovations necessary to leverage them—constitutes a major challenge for making use of advances in IT.

The dependence on models for rapid initial deployment is an especially striking aspect of this scenario. Aggregating the status information from FEMA, DHS, the Red Cross, DHHS, and DOD logistics and human resource systems is particularly difficult. Without a significant level of confidence in status information, managers cannot be sure that they know what impacts have occurred, what resources are needed, or what resources are responding. Models could also be valuable to responders and other on-the-scene personnel whose perspective might lead to insights not necessarily accessible to more distant emergency managers. However, this level of sharing is not typical. Also, modeling results are not often shared with lower-level personnel, who might be able to act more efficiently if they were. Finding ways to give more people access to modeling systems could yield significant benefits.

# B

# Review of Interoperability Initiatives

A number of initiatives and programs at the federal, state, and local levels have worked to improve interoperability. Some 60 agencies and programs deal with various aspects of interoperable communications and are spread throughout the federal government, state, regional, and local agencies as well as public safety associations.

One major federal program, started in early 2001, was the National Institute of Justice's (NIJ's) Advanced Generation of Interoperability for Law Enforcement (AGILE) program, now the CommTech program. AGILE was established as a comprehensive program to concentrate on the need for improved public safety communications and information sharing to address the requirement for effective coordination, communication, and sharing of information among numerous criminal justice and public safety agencies. Three research areas identified as being of general interest to the AGILE program reflect a focus on both data and voice interoperability:[1]

- Information sharing—Address technological and policy obstacles to enable effective and efficient, on-demand sharing of database information in a regional area;

---

[1] National Institute of Justice, "Communications Interoperability and Information Sharing Technologies (AGILE R&D) Program Solicitation," Washington, D.C., June 2001, p. 2; available at http://www.ncjrs.gov/pdffiles1/nij/agile2001.pdf.

- High-bandwidth communications—Ability to transmit, from stationary and mobile platforms, high-bandwidth information, such as still images and near real-time video, using common standards and open architecture techniques;

- Voice communications—Ability to overcome disparate bands, frequencies, and waveform generation techniques to enable regional voice communications interoperability for day-to-day and mutual aid situations.

In October 2001, the NIJ sponsored the National Public Safety Wireless Interoperability Forum. The goals of the forum were to raise public safety wireless interoperability to the national level and to give forum participants the opportunity to develop a list of actions that could be taken to overcome the policy barriers to improving public safety wireless communications.[2] The forum's success led to the creation of the National Task Force on Interoperability (NTFI).

In February 2003, NTFI—a task force comprising members from 18 national associations, state and local elected and appointed officials, and public safety officials—issued a guide for public officials entitled "Why Can't We Talk?" that reiterated and extended the work done by the Public Safety Wireless Advisory Committee. The NTFI guide states that the inability of public safety officials to communicate with one another "threatens the public's safety and often results in unnecessary loss of lives and property."[3] It does note the importance of data communications interoperability, particularly the need for interagency planning and coordination to achieve it. However, the primary focus is voice communications, and it is emphasized that a lack of adequate spectrum for public safety communications is one of the major barriers to interoperable communications.

Although it continues to receive considerable attention, adequate spectrum is not the only (or even the most important) requirement for interoperable communications. The definition of public safety wireless communications interoperability developed by the NTFI and refined by SAFECOM, a program of the Department of Homeland Security's (DHS's) Office for Interoperability and Compatibility, describes interoperability as "the ability of emergency response officials to share information via

---

[2]See the National Institute of Justice's information on the National Task Force on Interoperability at http://www.ojp.usdoj.gov/nij/topics/commtech/ntfi/welcome.html.

[3]National Task Force on Interoperability, "Why Can't We Talk?: Working Together to Bridge the Communications Gap, a Guide for Public Officials," February 2003, pp. 15-21; available at http://www.safecomprogram.gov/SAFECOM/library/interoperabilitybasics/1159_nationaltask.htm.

voice and data signals on demand, in real time, when needed, and as authorized."[4]

The NTFI guide identifies five key barriers to public safety communications interoperability:

- Incompatible and aging communications equipment,
- Limited and fragmented budget cycles and funding,
- Limited and fragmented planning and coordination,
- Limited and fragmented radio spectrum, and
- Limited equipment standards.

The Summit on Interoperable Communications for Public Safety, held in June 2003, was a joint effort among the National Institute of Standards and Technology (NIST), DHS's SAFECOM program, and NIJ's AGILE program. The summit, which brought together a variety of participants from federal, state, and national programs created to assist public safety practitioners, produced a briefing book listing the various programs and agencies at all levels of government involved in ongoing interoperability efforts.[5] The summit was intended as the initial step in familiarizing key interoperability players with the work being done by others so that mutually beneficial coordination and collaboration among the various technical programs could be established. The summit was also meant to provide insight into where additional federal resources might be warranted and to help stakeholders maximize the limited resources available across all government levels by leveraging program successes and developing standards, approaches, products, and services for the benefit of all.

One of the clear messages from the summit was that interoperability should be built from the bottom up. That is, interoperation of communications must be built starting at the state, local, and regional levels with guidance and support coming from the federal level. Under this framework, federal programs such as DHS's SAFECOM and NIJ's CommTech see their role as assisting state and local law enforcement agencies to communicate effectively and efficiently with one another across agency and jurisdictional boundaries.[6]

---

[4]See http://www.safecomprogram.gov/SAFECOM/interoperability/default.htm.

[5]Summit on Interoperable Communications for Public Safety, "Briefing Book of Public Safety Related Groups and Programs on Interoperable Communications and Information Sharing," NIST, Gaithersburg, Md., June 26-27, 2003, 85 pages. The list in the briefing book is the basis for the estimation that more than 60 programs are involved in various aspects of interoperability.

[6]The CommTech Web site notes this focus, stressing its role as a facilitator; see http://www.ojp.usdoj.gov/nij/topics/commtech/welcome.html.

The mission of the DHS S&T Directorate's Office for Interoperability and Compatibility (OIC) is to serve as the overarching program within DHS to strengthen and integrate interoperability efforts that improve public safety preparedness and response at all levels. OIC responsibilities include the following:

- Support the creation of interoperability standards;
- Establish a comprehensive research, development, testing, and evaluation program;
- Identify and certify all DHS programs that involve interoperability issues;
- Integrate coordinated grant guidance;
- Oversee the development and implementation technical assistance;
- Conduct pilot demonstrations of technology;
- Create an interagency interoperability coordination council; and
- Establish an effective outreach program.

OIC created the umbrella program SAFECOM to coordinate the efforts of public safety agencies at all levels of government in order to improve public safety response through more effective, efficient, interoperable wireless communications. SAFECOM has developed a number of tools and documents for public safety officials, including the Interoperability Continuum framework for addressing critical wireless interoperability elements; the Statewide Communications Interoperability Planning (SCIP) methodology; grant guidance, which includes guidelines for implementing a wireless communication system; and, in conjunction with the NIJ, a statement of requirements (SoR) for interoperability.[7]

Recent activity undertaken to carry out that mission includes the following:

- Creation of the Federal Interagency Coordination Council (FICC) to coordinate funding, technical assistance, standards development, and regulations affecting communications and interoperability across the federal government;
- Publication of a statement of requirements that defines, for the first time, what it will take to achieve full interoperability and provides industry requirements against which to map product capabilities;
- Issuance of a request for input from the public safety community

---

[7]Details on the SAFECOM program are available at http://www.safecomprogram.gov/SAFECOM.

in the form of a survey intended to help DHS identify and define specific projects to improve emergency communications;
- Initiation of an effort to accelerate the development of critical standards for interoperability;
- Creation of a grant guidance document that has been used by FEMA, Community Oriented Policing Services, and Office of Domestic Preparedness state block grant programs to promote interoperability improvement efforts;
- Establishment of a task force with the Federal Communications Commission to consider spectrum and regulatory issues that can strengthen public safety interoperability;
- Creation of the SCIP methodology for developing statewide communications plans;
- Release of a request for information to industry that netted more than 150 responses; and
- Work with the public safety community to develop a governance document that defines both how SAFECOM will operate and how participating agencies will work within that framework.

A number of states and regions have undertaken a variety of initiatives to achieve interoperable wireless communications among agencies in their areas. Virginia worked with SAFECOM and the NIJ to develop a strategic plan for improving statewide interoperable communications. The lessons from this effort were used by SAFECOM to create the SCIP methodology.

The NTFI identified a number of state and regional wireless voice communications interoperability efforts, including the Capital Wireless Integrated Network (CapWIN), which is a partnership between the state of Maryland, the Commonwealth of Virginia, and the District of Columbia to develop an interoperable first responder data communication and information sharing network; the state of South Dakota radio system, which includes everything from the wireless infrastructure to radios used in state-wide communications; and the state of Indiana radio system (Hoosier Safe-T), which provides statewide wireless infrastructure for voice communications. Other initiatives include the Minnesota metropolitan public safety radio system, public safety radio interoperability in Colorado, state of North Carolina interoperability initiatives, and the Utah communications agency network.[8] The nature of these efforts varies depending on the unique situation of the locality, region, or state.

---

[8]National Task Force on Interoperability, "Why Can't We Talk?: Working Together to Bridge the Communications Gap to Save Lives: A Guide for Public Officials, Supplemental Resource," February 2003, describes a number of state and regional interoperability case studies; available at http://www.ncjrs.gov/pdffiles1/nij/211079.pdf.

Efforts specifically addressing data communications interoperability have been the focus of considerable effort within the public safety community. FEMA's Disaster Management e-Gov initiative and its Disaster Management Interoperability Services (DMIS)[9] provide a set of services and software tools to public safety agencies for enabling responders to share information seamlessly between organizations. The Global Justice XML Data Model (GJXDM) and the derivative DHS National Information Exchange Model (NIEM) are intended as data reference models for the exchange of information within the justice and public safety communities.[10] Overarching efforts such as the Federal Enterprise Architecture and the e-Authentication e-Gov program also address information sharing and data exchange at the federal level.[11]

Industry-led efforts at data communications interoperability include the IEEE 1512 Standards Program, oriented to transportation incident management, and the IEEE 1451 Program for Sensor Integration Standards.[12] International efforts that span the government, private, and open-source communities include the Organization for the Advancement of Structured Information Standards Emergency Management Technical Committee, developers of the Common Alerting Protocol and the Emergency Data Exchange Language; and the Open GIS Consortium, creators of Geospatial Markup Language and of the Web Map Service and Web Feature Service specifications, among others.

Two significant examples of regional cooperation to provide wireless data interoperability are (1) the CapWIN wireless integrated mobile data communications network involving 41 federal, state, and local agencies serving the U.S. Capital Region (Washington, D.C., Maryland, and Virginia);[13] and (2) the Automated Regional Justice Information System (ARJIS) in the greater San Diego region, including San Diego and Imperial Counties in California, used by 71 local, state, and federal agencies to provide wireless data access to critical information in the field.[14]

---

[9]See http://www.cmi-services.org/.
[10]More information on GJXDM is available at http://it.ojp.gov/jxdm; further information on NIEM is available at http://niem.gov/.
[11]See http://www.whitehouse.gov/omb/egov/a-1-fea.html.
[12]For more information on IEEE 1512 and IEEE 1451, see http://grouper.ieee.org/groups/scc32/imwg/ and http://grouper.ieee.org/groups/1451/0/, respectively.
[13]Additional information on CapWIN is available at http://www.capwin.org/index.cfm.
[14]Additional information on ARJIS is available at http://www.arjis.org.

# C

# Workshop Agenda

**WEDNESDAY, JUNE 22, 2005**
**WASHINGTON, D.C.**

10:00–10:30 a.m.     **Welcome to the Workshop**

*Jon Eisenberg,* Study Director and Senior Program Officer, Computer Science and Telecommunications Board/National Research Council (NRC)
*Charles Brownstein,* Director, Computer Science and Telecommunications Board/NRC
*Ramesh Rao,* Chair, NRC Committee on Using Information Technology to Enhance Disaster Management—Improving the Use of Information Technology in Disaster Preparedness, Response, and Recovery; and Professor, University of California, San Diego
*Barry West,* Chief Information Officer/Director of Information Technology Services Division, Federal Emergency Management Agency (FEMA)

*Topic 1: The Critical and Evolving Role of Information and Communication Technologies (ICTs) in Disaster Management*

10:30 a.m–12:00 p.m.  **Context for and Visions of the Future, Grand Challenges for ICTs in Disaster Management**
*15-minute presentations by each panelist, followed by 15 minutes of Q&A*

*John Harrald*, Director, Institute for Crisis, Disaster, and Risk Management, George Washington University (*Moderator*)
*David G. Boyd,* Director, Office for Interoperability and Compatibility, Science and Technology Directorate, U.S. Department of Homeland Security
*Helen Wood*, Senior Advisor, National Oceanic and Atmospheric Administration Satellites and Information Service, and Chair, National Science and Technology Council's Subcommittee on Disaster Reduction
*Jack Potter,* Director, Emergency Medical Services, Valley Health, and Vice Chair, COMCARE Board of Directors
*Peter Miller*, Program Manager, Mission Support Office, Homeland Security Advanced Research Project Agency, Science and Technology Directorate, U.S. Department of Homeland Security

How might we better manage future crises? What impact will major technology trends have on how disaster management is handled in the future? How can we leverage commercial technology cost-performance curves yet meet special requirements? What new technologies, approaches, and policies would help, and what research directions are promising?

12:00–12:45 p.m.   Lunch

| | |
|---|---|
| 12:45–2:00 p.m. | **Perspectives on the Current State of the Art: ICTs in Disaster Management Practice**<br>*15-minute presentations by each panelist, followed by 15 minutes of Q&A*<br><br>*Ellis Stanley*, General Manager, Emergency Preparedness Department, City of Los Angeles, California (*Moderator*)<br>*Mark Deputy*, Senior IT Specialist and Urban Search and Rescue Team Assistant Task Force Leader, Montgomery County, Maryland<br>*William Maheu*, Executive Assistant Chief of Police, San Diego Police Department<br>*Robert Roth*, Fire Technology Specialist, U.S. Forest Service, U.S. Department of Agriculture<br><br>What lessons have been learned from past crisis and exercises? What gaps in IT capabilities are evident, and how might they be addressed? Where is the state of the art not sufficient to meet our needs? |
| 2:00–2:15 p.m. | Break |
| 2:15–3:30 p.m. | **Emerging Applications and Other Drivers for ICTs in Disaster Management**<br>*15-minute presentations by each panelist, followed by 15 minutes of Q&A*<br><br>*William Maheu*, Executive Assistant Chief of Police, San Diego Police Department (*Moderator*)<br>*William Metz*, Director, Center for Integrated Emergency Preparedness, Argonne National Laboratory<br>*Jaime Gomezjurado*, Project Manager and Vice President, Business Development, Medical Emergency Response Network Research Project, Semandex Networks, Inc.<br>*Lois Clark McCoy*, President, National Institute for Urban Search and Rescue<br>*Peter Brooks*, Institute for Defense Analyses |

|  |  |
|---|---|
|  | What are especially demanding aspects of disaster management? What initiatives are underway to address them? What specific requirements are there from specific groups or application areas? What general lessons can be learned? |
| 3:30–3:45 p.m. | Break |

*Topic 2: Research Directions for ICTs in Disaster Management*

|  |  |
|---|---|
|  | What research areas have application to disaster management? How might leading-edge research be applied to disaster management? How should the research agenda be established and evolved? How can research results best be transitioned into deployed capabilities? What are the commonalities between commercial, civilian, and military capabilities and research and development activities, and how can technology advances and knowledge be transferred from one to the other? |
| 3:45–5:00 p.m. | **Current ICT Research Programs Related to Disaster Management**<br>*15-minute presentation by each panelist, followed by 15 minutes of Q&A*<br><br>*Peter Steenkiste*, Professor, Carnegie Mellon University (*Moderator*)<br>*Larry Brandt*, Program Manager, Digital Government, National Science Foundation<br>*Gary Ham*, Senior Research Scientist, Battelle Memorial Institute<br>*Earnest Paylor*, Program Director, Pacific Disaster Center and Senior Advisor for Interagency Programs, Office of the Special Assistant and NASA Liaison to the Assistant Secretary of Defense—Networks and Information Integration<br>*Pamela Sydelko*, Leader—Modeling, Simulation and Visualization Group, Decision and Information Sciences Division, Argonne National Laboratory |

5:00–6:00 p.m     **ICT Everywhere: Ubiquitous and Pervasive Mobile (Ad Hoc) Communications and Networking**
*15-minute presentation by each panelist, followed by 15 minutes of Q&A*

*Timothy Brown*, Associate Professor, University of Colorado, Boulder (*Moderator*)
*Richard Howard*, Research Professor, Wireless Information Network Laboratory, Rutgers University
*Scott Midkiff*, Professor, Virginia Polytechnic Institute and State University
*Mani Chandy*, Simon Ramo Professor of Computer Science, California Institute of Technology

**THURSDAY, JUNE 23, 2005**
**WASHINGTON, D.C.**

*Topic 2: Research Directions for ICTs in Disaster Management (continued)*

8:30–9:45 a.m.     **Identifying and Aggregating Useful Data— Information Integration and Fusion**
*15-minute presentation by each panelist, followed by 15 minutes of Q&A*

*Gio Wiederhold*, Professor (Emeritus), Stanford University (*Moderator*)
*Yigal Arens*, Director, Intelligent Systems Division, Information Sciences Institute, University of Southern California
*Craig Knoblock*, Research Associate Professor, University of Southern California
*Peter Scott*, Associate Professor, University of Buffalo
*Zachary Ives*, Assistant Professor, University of Pennsylvania

9:45–10:45 a.m. **Information Overload: Making Useful Data Actionable—Decision Support, Collaboration, Situational Awareness**
*15-minute presentations by each panelist, followed by 15 minutes of Q&A*

*Robert Neches*, Director, Distributed Scalable Systems Division, Information Sciences Institute, University of Southern California (*Moderator*)
*David Mendonça*, Assistant Professor, New Jersey Institute of Technology
*William Wallace*, Professor, Rensselaer Polytechnic Institute

10:45–11:00 a.m. Break

11:00 a.m–12:15 p.m. **Sensor Networks, Autonomous Devices, and Geographic Information Systems**
*15-minute presentations by each panelist, followed by 15 minutes of Q&A*

*Robin Murphy*, Professor, University of South Florida, and Director, Industry/University Cooperative Research Center on Safety Security Rescue (*Moderator*)
*Ayman Mosallam*, Professor, Civil and Environmental Engineering, University of California, Irvine
*Susan McGrath*, Associate Research Professor of Engineering, Dartmouth College
*David Kehrlein*, Senior Consultant, Environmental Science Research Institute; formerly GIS Manager, California Governor's Office of Emergency Services

12:15–1:00 p.m. Lunch

APPENDIX C

*Topic 3: Collaboration, Coordination, and Interoperability: Pressing Issues in a Need-to-Share World*

1:00–3:00 p.m.  **Current Initiatives, Technical and Organizational Obstacles, and Opportunities in ICT Interoperability**
*15-minute presentations by each panelist, followed by 15 minutes of Q&A*

*Art Botterell*, Contra Costa County, California, Office of the Sheriff (*Moderator*)
*Ellis Kitchen*, Chief Information Officer, Maryland Office of Information Technology, and Member, Interoperability and Integration Committee, National Association of State Chief Information Officers
*Steve Cooper*, Senior Vice President and Chief Information Officer, Red Cross
*Lloyd (Gene) Krase*, Administrator, Kansas Division of Emergency Management
*Otto Doll,* Commissioner, Bureau of Information and Telecommunications, South Dakota
*Dave Smith*, Implementation Director, Indiana Integrated Public Safety Commission
*Robert Fletcher*, President, Readiness Consulting Services, LLC, and Member, National Fire Protection Association NFPA 1600 Technical Committee

What policy and technical initiatives are currently underway to improve wireless interoperability across federal, state, and local levels? What results are anticipated, and over what time frames? How are data (including voice) generated, used, and shared across organizational boundaries today? What can be learned in terms of technology and practice from efforts to create systems capable of greater interoperability? How are standards helping or not helping? What kinds of interoperability are desired in the future? What technical, operational, economic, and policy challenges are likely to be unresolved, and merit further research? How do communication systems

|               |   |
|---------------|---|
| | relate to other information management systems related to disaster management? What obstacles to technology transition must be overcome? |
| 3:00–3:15 p.m. | Break |
| 3:15–4:30 p.m. | **Envisioning, Enabling, and Building Networks of the Future** |

*15-minute presentation by each panelist, followed by 15 minutes of Q&A*

*Nancy Jesuale*, President, NetCity Engineering (*Moderator*)
*Nader Moayeri*, Manager, Wireless Communications Technologies Group, National Institute of Standards and Technology
*James Morentz*, Vice President, Homeland Security Technology, and Director, Public Safety Integration Center, Science Applications International Corporation
*Chip Hines*, Program Manager, Disaster Management eGov Initiative, Office of the Chief Information Officer, Emergency Preparedness and Response/FEMA, Department of Homeland Security

What should communications and other IT capabilities look like in the future? How do wireless communications systems relate to the emerging broader architecture for public safety, national security, and disaster response? What are the implications of major information and communications technology trends for how we respond to crises? How can these opportunities better be exploited? What kinds of research, experimentation, and pilot programs would help?

4:30–4:45 p.m.   **Concluding Remarks**

*Ramesh Rao*, Chair, NRC Committee on Using Information Technology to Enhance Disaster Management

# D

# Speakers and Participants at Meetings and Site Visits

Although the briefers listed below provided much useful information of various kinds to the Committee on Using Information Technology to Enhance Disaster Management, they were not asked to endorse the conclusions or recommendations of this study, nor did they see the final draft of this report before its release.

### SEPTEMBER 20-21, 2005
### WASHINGTON, D.C.

**Briefers at Meetings:**

Alok Chaturvedi, Purdue University
Louise Comfort, University of Pittsburgh
Kenneth Mandl, Harvard Medical School
Charles Werner, Charlottesville (Virginia) Fire Department

**Site Visit:**

Science Applications International Corporation (SAIC) Public Safety Integration Center, hosted by James Morentz, SAIC

## DECEMBER 12-13, 2005
## SAN DIEGO, CALIFORNIA

**Briefers at Meetings:**

Mark Koro, Qualcomm
Leslie Lenert, University of California, San Diego
Bill Owens, Nortel Networks (retired)
Pam Scanlon, Automated Regional Justice Information System
Christian Sloane, University of California, San Diego Medical Center

**Site Visits:**

San Diego Police Department Disaster Operation Center
City of San Diego Emergency Operation Center
San Diego County Disaster Operation Center
San Diego SWAT Demonstration

## MAY 1, 2006
## WASHINGTON, D.C.

**Briefers at Meetings:**

Tom Coty, Homeland Security Institute
Felix Demicco, Office of Critical Infrastructure—Office of the Prosecutor, Morris County, New Jersey
Christopher Kojm, 9/11 Public Discourse Project and Deputy Director, 9/11 Commission
Dereck Orr, Public Safety Communications Systems, National Institute of Standards and Technology
Florence Reichenberg, Morris County, New Jersey, Chamber of Commerce
Havidán Rodriguez, Disaster Research Center, University of Delaware

# E

# Biographies of Committee Members and Staff

**COMMITTEE MEMBERS**

**Ramesh R. Rao,** *Chair,* is currently a professor in the Department of Electrical and Computer Engineering and director of the San Diego Division of the California Institute of Telecommunications and Information Technology (Calit2) at the University of California, San Diego (UCSD). His research interests include architectures, protocols, and performance analysis of wireless, wire line, and photonic networks for integrated multimedia services. Prior to his appointment as the director of the San Diego Division of Calit2, he served as director of the UCSD Center for Wireless Communications (CWC) and was the vice chair of Instructional Affairs in the Department of Electrical and Computer Engineering. Professor Rao did his undergraduate work at the Regional Engineering College of the University of Madras in Tiruchirapalli, India, obtaining a B.E. (honors) degree in electronics and communications in 1980. He did his graduate work at the University of Maryland, College Park, Maryland, receiving his M.S. in 1982 and his Ph.D. in 1984.

**Yigal Arens** is director of the Intelligent Systems Division of the University of Southern California's (USC) Information Sciences Institute. He is also co-director of the USC/Columbia University Digital Government Research Center (DGRC) and a research professor at USC's Daniel J. Epstein Department of Industrial and Systems Engineering. His primary research interests have been digital government, information integration, planning in the domain of information servers, knowledge representation, and human-machine communication. In 1983, he joined the faculty of USC's Computer Science Department. He joined USC's Information Sciences Institute (USC/ISI) in 1987, where he first worked on the Integrated Interfaces project, a multimedia presentation design system combining text, tables, maps, and other graphics. For almost 10 years he headed the Single Interface to Multiple Sources research group specializing in integration of heterogeneous databases and other information sources. Dr. Arens has been director of the Intelligent Systems Division, one of the largest artificial intelligence research laboratories in the United States, since 1999. Also, since 1999, he has been co-director of the DGRC. In 1999, together with two colleagues from ISI, he founded Fetch Technologies, a company that specializes in extracting data from Web sites. In 2002, he joined the Daniel J. Epstein Department of Industrial and Systems Engineering as a research professor. In 2003, Dr. Arens founded USC's Center for Research on Unexpected Events, which he headed for its first year. Dr. Arens also was a part of the National Research Council's Committee on Planning Meeting on Information Technology and the States: Public Policy and Public. He received his Ph.D. from the University of California at Berkeley.

**Art Botterell** is community warning system manager, Contra Costa County (California) Office of the Sheriff. He is an internationally recognized expert in emergency communications who has served on the front lines of some of the biggest national disasters in recent U.S. history. Former Federal Emergency Management Agency director James Lee Witt hailed him as a "national asset." He has served as a consultant to the Department of Homeland Security and a number of other state, federal, and international organizations. He led the development of the Common Alerting Protocol (CAP)—the first international standard format for all-hazard public warning across multiple media. An experienced analyst, broadcast and multimedia producer, writer, and manager, Mr. Botterell studies the ways that communities use information technology to manage the effects of sudden change.

**Timothy X. Brown** is an associate professor at the University of Colorado, Boulder. He received his B.S. in physics from Pennsylvania State University and his Ph.D. in electrical engineering from the California Institute of Technology in 1990, when he also joined the Jet Propulsion Laboratory. In 1992 he joined Bell Communications Research. Since 1995 he has held a joint appointment with Electrical Engineering and Interdisciplinary Telecommunications at the University of Colorado, Boulder. Dr. Brown's research interests include adaptive network control, machine learning, and wireless communications systems. His laboratory has developed extensive experience in the design, implementation, and testing of wireless networking protocols. He has published more than 50 papers in networking and wireless systems, is a recipient of the National Science Foundation's CAREER Award, and was named the Global Wireless Education Consortium Wireless Educator of 2003.

**John R. Harrald** is the director of the George Washington University (GWU) Institute for Crisis, Disaster, and Risk Management and a professor of engineering management in the GWU School of Engineering and Applied Science. He is a founding member, director, and immediate past president of the International Emergency Management Society. Dr. Harrald has been actively engaged in the fields of emergency, consequence, and crisis management and maritime safety and port security. He was the former director of the Disaster Recovery Institute (DRI) and served as the associate director of the National Ports and Waterways Institute for 10 years. Dr. Harrald was the principal investigator for maritime risk and crisis management studies in Prince William Sound, Alaska, the Port of New Orleans, and Washington state, and for earthquake vulnerability studies funded by the National Science Foundation and the American Red Cross. He has studied the response to the *Exxon Valdez* oil spill, the Loma Prieta earthquake, Hurricane Hugo, Hurricane Andrew, the Northridge earthquake, the 1999 Turkey earthquakes, and the September 11, 2001, terrorist attacks on the World Trade Center and the Pentagon. He has also written and published in the fields of crisis management, emergency management, management science, risk and vulnerability analysis, and maritime safety. He was a reviewer for the committee that produced *Information Technology, Research, Innovation, and E-Government*. Dr. Harrald received his B.S. in engineering from the U.S. Coast Guard Academy, an M.A.L.S. from Wesleyan University, an M.S. from the Massachusetts Institute of Technology where he was an Alfred P. Sloan Fellow, and an M.B.A. and a Ph.D. from Rensselaer Polytechnic Institute.

**Richard Howard** is a researcher at Wireless Information Network Laboratory (WINLAB) at Rutgers University. He is also a principal at Research Innovations, LLC, and the founder and senior vice president of technology at PnP Networks, a start-up company focused on applying artificial intelligence techniques to the problem of making computers and computer networks truly simple for people to use. Dr. Howard was formerly the wireless research vice president at Lucent Bell Laboratories, where he did research on wireless technology ranging from materials, components, packaging, antennas, modeling, analysis, communication theory, and integrated circuit design to systems-level projects such as fixed wireless loop and advanced cellular base stations. His work has emphasized multiple antennas, signal processing, and system performance from basic communication theory to field deployment. Dr. Howard's key achievements have included new theory (and practical demonstrations) for dramatically increasing wireless system capacities based on multiple antennas. Other achievements have included algorithms and a tool suite for optimization of cellular networks and application of advanced signal processing to linear power amplifiers for dramatic reductions in cost and size and improvements in efficiency. He received his Ph.D. in physics at Stanford University in 1977.

**Nancy Jesuale** has worked in local and state government since 1976 as a telecommunications strategic planner and has served as a director of public safety networks, telecommunications networking, and network operations. Ms. Jesuale is the president of NetCity Engineering, Inc. (NCE), a consulting practice dedicated to strategic planning and solution sets for government in public safety and fiber-optic telecommunications systems. Current clients of NCE include the city of Los Angeles; the District of Columbia; the state of Oregon; the city of Charlotte, North Carolina; and the Center for Wireless Network Security (WiNSeC). As program manager for public safety for the Center for WiNSeC at the Stevens Institute of Technology, Ms. Jesuale is responsible for establishing relationships, research programs, and public policy support. She has been an innovator in telecommunications strategies for local government since 1984. She is an appointee to the National Task Force on Interoperability and the Oregon State Interoperability Executive Committee; is a past chair of the Public Technology, Inc., Task Force on Information Technology and Telecommunications; has been the director of strategic planning for telecommunications for the city of Los Angeles; and has served on the Oregon Statewide Interoperability Executive Council.

**David Kehrlein,** now with the Environmental Science Research Institute (ESRI), was the geographical information systems (GIS) manager for the California Governor's Office of Emergency Services (OES) for more than 9 years. Before that he worked in the Forest and Rangeland Resource Assessment Program (FRRAP) of the California Department of Forestry and Fire Protection. Mr. Kehrlein was active on the governor's GIS task force in 1992. He is a past director of the California Geographic Information Association (CGIA), and he chairs the data standards committee. He was also chair of the Firefighting Resources of California Organized for Potential Emergencies (FIRESCOPE) GIS Specialist Group. He has organized response and recovery GIS support for 16 presidentially declared disasters, from incident-level response to decision support at the state and federal levels. His group at OES also deployed a response/training GIS trailer that is equipped with large-format plotting and scanning capabilities, a statewide GIS data repository, as well as satellite cell phone and a high-speed satellite Internet downlinking capability. Mr. Kehrlein received his B.A., graduating with honors in geography from California State University, Sacramento.

**William Maheu** is chief of operations of the San Diego, California, Police Department. Mr. Maheu has been a member of the police department for 23 years. He is currently in charge of Child Abuse, Domestic Violence, Sex Crimes, Vice Operations, Mid City Division, Southeastern Division, Southern Division, Records, Property and various other programs. During his tenure with the department, he has had many assignments, including commanding officer of field operations/special resources, executive lieutenant of the Special Weapons and Tactics Team, special projects/long-range planning lieutenant and narcotics sergeant. He has also been involved in several major projects, including the 2003 Super Bowl, the Republican National Convention, the Presidential Debate, development of the Psychiatric Emergency Response Team, and the development of the Homeless Outreach Team. Mr. Maheu graduated from the University of San Diego in 1983 with a B.A. in psychology.

**Robin R. Murphy** is a professor in the Computer Science and Engineering Department at the University of South Florida, with a joint appointment in cognitive and neural sciences in the Department of Psychology. She is an associate editor for the Institute of Electrical and Electronics Engineers *Intelligent Systems* and a member of the 1998-1999 Defense Science Study Group and is currently a member of the U.S. Air Force Scientific Advisory Board and the Defense Advanced Research Projects Agency Innovative Space-based radar Antenna Technology (ISAT). She recently served on the Department of Defense's Air Platforms FY2004 Technology Area Review. In addition, she is also a member of the board of directors for Continental Divide Robotics, which provides the Global Positioning System and intelligent-agent software for tracking parolees. From 1992 to 1998, she was an assistant professor in the Department of Mathematical and Computer Sciences at the Colorado School of Mines. Dr. Murphy joined the University of South Florida (USF) in 1998, and in January 2002 she became director of the Center for Robot-Assisted Search and Rescue (CRASAR). In March 2003, she helped start the Industry/University Cooperative Research Center on Safety Security Rescue (SSR-RC) with the University of Minnesota and is the overall director. She leads the CRASAR rescue robot response team, the only such team in the world, and is a technical search specialist with Florida Task Force 3. Since 1995, she has focused on Urban Search and Rescue (USAR) as the test domain for her research, leading to her participation in the first known use of robots for urban search and rescue at the World Trade Center disaster. Her USAR robotics work has earned a National Institute for Urban Search and Rescue Eagle award, and she serves on the executive board of the National Institute for Urban Search and Rescue. She has also won a USF Outstanding Faculty Research Achievement Award (2003) and received the Honor Society of Phi Kappa Phi, USF Chapter, Artist and Scholar of the Year Award (2004). Prior to graduate work, Dr. Murphy worked in the process control industry as a software project engineer. She served as a committee member for the Army Unmanned Ground Vehicle Technology Committee of the National Research Council. She received a B.M.E. in mechanical engineering and an M.S. and a Ph.D. in computer science (with a minor in computer integrated manufacturing systems) in 1980, 1989, and 1992, respectively, from Georgia Institute of Technology, where she was a Rockwell International Doctoral Fellow.

**Robert Neches** is the director of the Information Sciences Institute's Distributed Scalable Systems Division and a research faculty member of the University of Southern California's Computer Science Department. He received his Ph.D. from Carnegie Mellon University in 1981 for work in machine learning, spent a year at the University of Pittsburgh's Learning Research and Development Center, and has been at USC ISI since 1982 (with the exception of service at the Defense Advanced Research Projects Agency [DARPA] during 1994 to 1997). His personal interests span control and coordination in distributed systems, collaboration and visualization aids for information management, and system-of-systems frameworks for information integration. The Distributed Scalable Systems Division looks at the full range of issues bearing on organizations' gathering of information, assessing it, making decisions, reconciling issues, and effecting resulting actions. Research within the division addresses distributed software systems engineering, information management, intelligent human-computer interaction, computer-supported cooperative work, resource management, and decision support. Applications within the division include all levels of command and control, crisis management, intelligence analysis, logistics, design and manufacturing, and space applications.

**Masanobu Shinozuka** is a Distinguished Professor and chair of the Department of Civil and Environmental Engineering at the University of California, Irvine, and Norman Sollenberger Professor Emeritus of Civil Engineering at Princeton University. He is a member of the National

Academy of Engineering. His research activities involve random vibration, reliability of structural systems, structural dynamics, structural control, continuum mechanics, and infrastructure systems including lifeline networks. In particular, his pioneering and original research on digital simulation of stochastic waves is noteworthy. He has more than 500 publications in refereed journals and proceedings of national and international conferences in mechanics, structural engineering, and natural/human-made disaster mitigation. His contribution to these areas was recognized in terms of a number of prestigious awards, such as Newmark, Freudenthal, and Von Karman Medals from the American Society of Civil Engineers, of which he is an honorary member. Professor Shinozuka's recent research deals with the detection of damage and its locations within a network of utility and highway transportation systems under natural and human-made disturbances. In this regard, his most recent effort, under the sponsorship of the National Science Foundation, focuses on the development of energy-efficient and self-powered sensor networks and wireless data transmission systems that can be applied to real-time diagnosis of these systems after serious security breaches. He has a long history of working relationships with engineers and management at the California Department of Transportation, the Los Angeles Department of Water and Power, and Memphis Light, Gas and Water, and more recently with Southern California Edison to estimate the seismic performance of their systems. He also served as president and executive vice president of the International Association of Structural Safety and Reliability. Professor Shinozuka received his Ph.D. from Columbia University from the Department of Civil Engineering and Engineering Mechanics in 1960 and an M.S. in civil engineering (1955) and a B.S. (1953) from Kyoto University.

**Ellis Stanley** is the general manager for the Emergency Preparedness Department for the city of Los Angeles. Currently he serves as an adviser to the Multidisciplinary Center for Earthquake Engineering Research and is a member of the center's Industry Advisory Board; he also chairs the Metro Emergency Manager's Forum of the International Association of Emergency Managers. He is vice president for the public sector of the Business and Industry Council on Emergency Preparedness and Planning and is on the Emergency Services Committee of the American Red Cross, Los Angeles Chapter. The City Council has also appointed him to the Emergency Preparedness Commission for the county and city of Los Angeles, and he is a member of the city's Emergency Operations Board. Mr. Stanley was recently appointed to the board of directors of the National Institute of Urban Search and Rescue. He was the director of the Atlanta-Fulton County Emergency Management Agency and has been the director of an emergency management program for the city of Durham and Durham County, North Carolina, and Brunswick County, North Carolina. He also served as a county fire marshal, fire and rescue commissioner, and county safety officer; as president of the International Association of Emergency Managers, the American Society of Professional Emergency Planners, the National Defense Transportation Association, and the Metropolitan Atlanta Chapter of the National Forum for Black Public Administrators; and as vice chair of the Association of Contingency Planners. He also chaired the Certified Emergency Managers Certification Commission. Mr. Stanley is a 1973 graduate of the University of North Carolina at Chapel Hill with a degree in political science.

**Peter Steenkiste** is a professor of computer science and of the Electrical and Computer Engineering Department at Carnegie Mellon University (CMU). He is currently on the editorial board of *IEEE/ACM Transactions on Networking, Cluster Computing,* and the *Journal of Grid Computing*. His current research is in the areas of network services and pervasive computing. He is currently working on the Darwin project and is also active in pervasive computing in the context of the CMU Aura project. Professor Steenkiste's other research interests are in the areas of networking and distributed computing. While at CMU, he worked on Nectar, the first workstation clusters built around a high-performance, switch-based local area network. He is a member of the Association for Computing Machinery (ACM) and a senior member of the Institute of Electrical and Electronics Engineers (IEEE). He has been on a number of program committees and was co-chair for the OPENSIG'99 workshop and the Eighth International Workshop on Quality of Service. He was also program chair for HPDC'2000 and general co-chair for ACM SIGCOMM'02. He was an associate editor for *IEEE Transactions on Parallel and Distributed Systems* (1998-1999). He received a degree in electrical engineering from the University of Gent in Belgium in 1982, and his M.S. and Ph.D. degrees in electrical engineering from Stanford University in 1983 and 1987, respectively.

**Gio Wiederhold** is an emeritus professor of computer science at Stanford University, with courtesy appointments in medicine and electrical engineering. His current research includes privacy protection in collaborative settings, large-scale software composition, access to simulations to augment decision-making capabilities for information systems, and developing algebra over ontologies. Prior to his academic career, he spent 16 years in the software industry. His career followed computer technologies, starting with numerical analysis applied to rocket fuel, FORTRAN and PL/1 compilers, real-time data acquisition, and a time-oriented database system; eventually he became a corporate software architect. He has been elected a fellow of the American College of Medical Informatics, the IEEE, and the ACM. He spent 1991 through 1994 as the program manager for knowledge-based systems at DARPA in Washington, D.C. He has been an editor and editor-in-chief of several IEEE and ACM publications. Professor Wiederhold served as a reviewer for several CSTB reports, including *Information Technology, Research, Innovation, and E-Government; Youth, Pornography, and the Internet; Technical, Business, and Legal Dimensions of Protecting Children from Pornography on the Internet: Proceedings of a Workshop; Non-technical Strategies to Reduce Children's Exposure to Inappropriate Material on the Internet: Summary of a Workshop; Review of the FBI's Trilogy Information Technology Modernization Program*; and a letter report to the FBI. He received a degree in aeronautical engineering in Holland in 1957 and a Ph.D. in medical information science from the University of California at San Francisco in 1976.

## STAFF

**Jon Eisenberg** is director of the Computer Science and Telecommunications Board (CSTB) of the National Research Council. At CSTB, he has been study director for a diverse body of work, including a series of studies exploring Internet and broadband policy and networking and communications technologies. Current studies include an examination of emerging wireless technologies and spectrum policy and a study of how to use information technologies to enhance disaster management. From 1995 through 1997 he was an American Academy of Arts and Science, Engineering, and Diplomacy Fellow at the U.S. Agency for International Development, where he worked on environmental management, technology transfer, and information and telecommunications policy issues. He received his Ph.D. in physics from the University of Washington in 1996 and a B.S. in physics with honors from the University of Massachusetts at Amherst in 1988.

**Ted Schmitt** is a program officer for the Computer Science and Telecommunications Board of the National Research Council. In addition to the present study, he is currently involved in the CSTB project on providing a comprehensive exploration of cybersecurity. Before serving with CSTB, Mr. Schmitt was involved in the development of the digital publishing industry and played an active role in various related standards groups. Prior to that, he served as technical director at a number of small technology companies in Germany, Sweden, and the United States. He started his career in 1984 as a software engineer for IBM, earning two patents and several technical achievement awards. Mr. Schmitt received an M.A. in international science and technology policy from George Washington University. He received a B.S. in electrical engineering in 1984 and a B.A. in German in 1997 from Purdue University and studied at the Universität Hamburg, Germany.

**Jennifer M. Bishop,** program associate, recently left the staff of the Computer Science and Telecommunications Board of the National Research Council. She was involved in several studies, including Telecommunications Research and Development, Policy Consequences and Legal/Ethical Implications of Offensive Information Warfare, and Assessing the Information Technology Research and Development Ecosystem. She also maintained CSTB's databases; managed the CSTB Web site; produced *Update*, the CSTB newsletter; and designed book covers and promotional materials. Prior to serving with CSTB, Ms. Bishop worked for the city of Ithaca, New York, coordinating the Police Department's transition to a new SQL-based time accrual and scheduling application, a project that grew out of her experience with maintaining the police records databases. Her other work experience includes designing customized hospitality industry performance reports for Smith Travel Research, and freelance publication design. She is interested in the social and cultural impacts of information technology, including researching and developing effective information design for education and lifelong learning. In her spare time, Ms. Bishop is a visual artist working in oil and mixed media. She holds a B.F.A. from Cornell University's College of Architecture, Art, and Planning.

**David Padgham,** associate program officer at the Computer Science and Telecommunications Board of the National Research Council, is currently involved in studies investigating dependable software, health care informatics, computing performance, and forensics. He rejoined CSTB in the spring of 2006 following nearly 2 years as a policy analyst in the Association for Computing Machinery's (ACM's) Washington, D.C., Office of Public Policy, where he worked closely with that organization's public policy committee, USACM. Previously, he spent nearly 6 years with CSTB, working on—among other things—the studies that produced *Trust in Cyberspace; Funding a Revolution; Broadband: Bringing Home the Bits; LC21: A Digital Strategy for the Library of Congress;* and *The Internet's Coming of Age.* He holds a master's degree in library and information science (2001) from the Catholic University of America in Washington, D.C., and a B.A. (1996) in English from Warren Wilson College in Asheville, North Carolina.

**Gloria Westbrook** recently left the staff of the Computer Science and Telecommunications Board, where she was a senior project assistant. She previously served as the executive assistant to the directors of the Office of Youth Programs and the Youth Opportunity Grant Program at the D.C. Department of Employment Services (DOES). In 2003, Ms. Westbrook was selected to lead a team that successfully administered a $4 million summer youth employment program that registered more than 5,000 District youth. In addition, Ms. Westbrook also served as the executive assistant to the director of the DOES and served as his liaison to the District of Columbia's mayor and his cabinet, council members, and members of Congress. While serving in the director's office Ms. Westbrook received the Meritorious Service Award and the Workforce Development Administrator's Award of Appreciation for Dedication of Service. She also became a member of the National Association of Executive Secretaries & Administrative Assistants. Ms. Westbrook attended Duke Ellington School of the Performing Arts in ballet and went on to further her dance education at the University of the Arts in Philadelphia.